A Handbook on the Requirements for Electrical Installations

The IEE Wiring Regulations 16th Edition

Based on BS 7671: 1992 incorporating
Amendment 2: December 1997

Published jointly by

ESCA House, 34 Palace Court,
London W2 4HY
Tel: 0171-313-4800

Bush House, Bush Estate,
Midlothian, EH26 0SB
Tel: 0131-445 5577

National Inspection Council for
Electrical Installation Contracting

Vintage House, 36/37 Albert
Embankment, London SE1 7UJ
Tel: 0171-582-7746

Blackwell
Science

Blackwell Science Ltd
Editorial Offices:
Osney Mead, Oxford OX2 0EL
25 John Street, London WC1N 2BL
23 Ainslie Place, Edinburgh EH3 6AJ
350 Main Street, Malden
MA 02148 5018, USA
54 University Street, Carlton
Victoria 3053, Australia
10, rue Casimir Delavigne
75006 Paris, France

Blackwell Wissenschafts-Verlag GmbH
Kurfürstendamm 57
10707 Berlin, Germany

Blackwell Science KK
MG Kodenmacho Building
7–10 Kodenmacho Nihombashi
Chuo-ku, Tokyo 104, Japan

First published by Blackwell Scientific Publications under the title *A Handbook on the 16th Edition of the IEE Regulations for Electrical Installations* 1991
Reprinted 1991 (three times), 1992 (twice)
First edition revised published by Blackwell Science 1995
Reprinted 1996
Second edition published 1998
Reprinted 1999

Printed and bound in Great Britain
by MPG Books Ltd, Bodmin, Cornwall

DISTRIBUTORS

Marston Book Services Ltd
PO Box 269
Abingdon
Oxon OX14 4YN
(*Orders:* Tel: 01235 465500
Fax: 01235 465555)

USA
Blackwell Science, Inc.
Commerce Place
350 Main Street
Malden, MA 02148 5018
(*Orders:* Tel: 800 759 6102
781 388 8250
Fax: 781 388 8255)

Canada
Login Brothers Book Company
324 Saulteaux Crescent
Winnipeg, Manitoba R3J 3T2
(*Orders:* Tel: 204 224 4068)

Australia
Blackwell Science Pty Ltd
54 University Street
Carlton, Victoria 3053
(*Orders:* Tel: 03 9347 0300
Fax: 03 9347 5001)

A catalogue record for this title is available from the British Library

ISBN 0-632-04952-9

For further information on Blackwell Science, visit our Website
www.blackwell-science.com

This publication was created using Ventura Publisher™

This edition technical authorship by John Peacock, on behalf of: The Electrical Contractors' Association 0171-313 4800; Select 0131-445 5577; The National Inspection Council for Electrical Installation Contracting 0171-582 7746.

This edition page layout by Archetype Information Technology Ltd, Condicote and Camplong d'Aude
Web site: http:www.archetype-it.com

Contents

Using the Handbook

This Handbook is arranged in six parts, including an index.

PART A

Introduction and Plan of BS 7671: 1992.

PART B

Index charts showing the arrangement of each chapter, accompanied where necessary by Points of Special Note. These charts also serve as an outline index to enable the reader to quickly identify Regulations relevant to his needs.

PART C

Topic charts, in which particular topics, e.g. isolation and switching, are presented as a co-ordinated reference to all the Regulations relevant to the subject, irrespective of the chapter in which they appear. Each topic chart is followed by supporting text, and is in one of two forms:

(i) decision charts, where the reader is led to the decision appropriate to his requirements; and

(ii) information charts and diagrams, where the reader is shown the access to the information for selecting the method or equipment most suitable.

PART D

Other illustrations for particular applications, indicating the requirements in visual form, together with explanatory notes.

PART E

Other useful sources of further information.

PART F

Index

NOTES

1. Parts B, C and D are cross-referenced as necessary.

2. It is anticipated that Parts A, B, C and E will be of greatest interest to the designer, whereas Parts C and D may be of more interest to the installer. The whole Handbook will be of use to college lecturers and students who require to study the Regulations in depth.

3. Regulation numbers throughout the book refer to BS 7671: 1992.

The co-operating bodies gratefully acknowledge the permission of the Institution of Electrical Engineers to quote from BS 7671: 1992.

Summary of Illustrations and Diagrams

A

Introduction and Plan of BS 7671: 1992

Introduction and Plan of BS 7671:1992 Including Amendments No 1 December 1994 and No 2 December 1997

General

After fourteen editions of the Regulations for the Electrical Equipment of Buildings had been produced by the Institution of Electrical Engineers over a period of a hundred years, the 15th Edition – retitled Regulations for Electrical Installations, was introduced in 1981. After several sets of Amendments to the Regulations, various alterations to British Standards and Codes of Practice, and changes in legislation, a 16th Edition was published in May 1991. In October of 1992 the 16th Edition became a British Standard 7671:1992.

This edition of the Handbook is a revised version updated to reflect the changes to BS 7671:1992 arising from amendments No 1 and 2.

Many references are made in the Regulations to British Standards and Codes of Practice. Several aspects of electrical installation work, however, are specifically excluded from the Regulations, other than for their interface at power supplies, e.g. Emergency Lighting and Fire Alarms. Lightning protection and work in hazardous atmospheres are also excluded. In these areas, therefore, the relevant Standards and Codes provide authoritative information and also refer to the relevant Wiring Regulations, they should be readily available to designers and specifiers. Other aspects, e.g. work in mines and quarries, are the subject of separate legislation, as are explosives factories. Health and Safety legislation enters into almost every activity: designers and others should be aware of the wide range of informative publications produced by HSE.

The Regulations are divided into distinct parts: a statement of the hazards inherent in the use of electricity (danger from electric shock, fire, burns and injury from mechanical movement), a description of the various methods of combating these, the application of these methods, and Appendices some of which contain guidance on standard arrangements which will satisfy the Regulations in certain circumstances. The current-carrying capacity tables are also contained in the Appendices.

It will be seen that BS 7671:1992 cannot be used, or even read easily, in serial form. It is necessary to follow a topic through various parts, chapters and sections, in order to achieve understanding.

Recognizing that Designers, Installers, Inspectors and Testers need a minimum accepted standard there are Guidance Notes published by the IEE which should be obtained and used in conjunction with a study of BS 7671:1992.

This Handbook is intended to guide the designer and installer through BS 7671:1992 and amendments Nos. 1 & 2 showing how each part is related to the whole, and to give practical guidance on how to approach new installations, extensions to existing installations, and the more extensive testing and inspection which is required.

Amendment Number 1

This was published in December 1994 to become effective on 1st July 1995. In the main the amendments have been made to take account of EU CENELEC changes including the changes to the nominal supply voltage from 240 volts to 230 volts effective from 1st July 1995. Further amendments were made to align with CENELEC harmonisation document 384 and to correct grammar and punctuation errors in the previous issue.

Significant changes come into the sections relating to isolation and switching, protective extra-low voltage (PELV), swimming pools, electircal installations in caravans and for caravan parks (reversion to the term parks in place of sites as in the original issue). Extensive revision and updating of the appendices has taken place particularly as they relate to external influences and the forms of completion and inspection.

Amendment Number 2

Like amendment number 1 this second amendment enhances some of the definitions and adds some new ones. Low voltage circuits are replaced by Band II circuits and Extra Low voltage circuits by Band I circuits. Many of the other amendments are to take account of EU CENELEC changes.

A significant change is the renumbering of section 551 as 555 so that 551 is used for the addition of a section dealing with Generating Sets.

Amendments have been made to Inspecting and Testing Certification and Reporting.

Within this handbook only amendments affecting regulations have been identified. Where there is a correction for punctuation or grammar, or a change to a BS number these have just been included in the text. Identification of changes in the text is by underlining.

Arrangement of BS 7671:1992

BS 7671:1992 consists of Seven Parts, within each of which there are Chapters on the general subject, and Sections on particular aspects. Since the main objective has been to bring international accord to the work of electrical installation, future amendments will cover progress in the field. The division into Parts, Chapters and Sections is intended to facilitate the inclusion of such amendments and extensions when agreed.

The numbering system used for individual regulations in BS 7671: 1992 is significant, and assists in cross referencing.

First digit	signifies the PART in which the Regulation is to be found
Second digit	signifies the CHAPTER of the Part
Third digit	signifies the SECTION of that Chapter

Subsequent digits identify the particular Regulation number in that section. For example, Regulation 522 - 03 - 01 is to be found in:

Part 5	which deals with Selection and Erection of Equipment
Chapter 2	which deals with Selection and Erection of Wiring Systems
Section 2	dealing with external influences
Sub-section 03	which deals with the presence of water (AD) or high humidity (AB)

This relationship is shown diagrammatically in Figure A1, and it is suggested that oral reference to a Regulation should be made thus:

FIVE TWO TWO DASH ZERO THREE DASH ZERO ONE

This enables the listener to clearly identify the Part, Chapter, Section and Regulation.

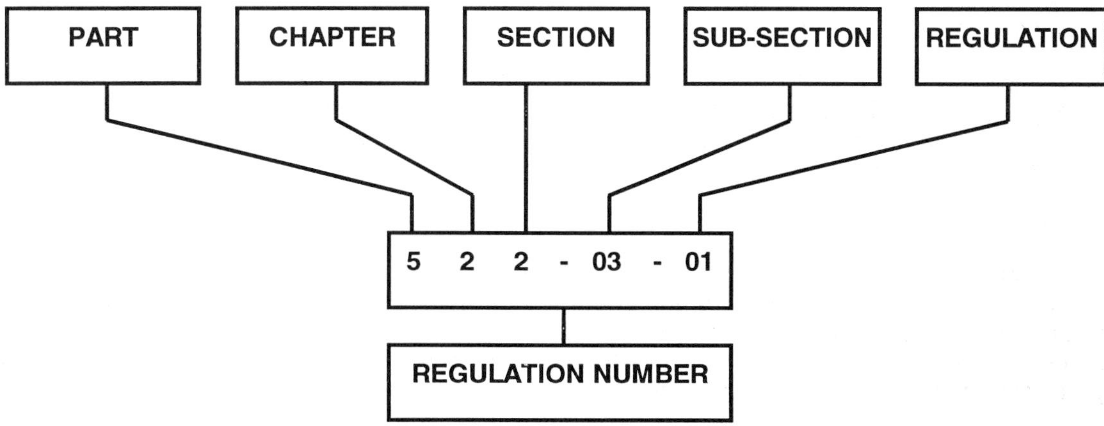

Figure A1: Derivation of Regulation Numbers

The plan of BS 7671:1992 is shown in two different ways in Figures A2 and A3:

(i) Figure A2 shows the plan by subject headings.

(ii) Figure A3 shows the interdependence of the various parts.

It will be seen that in Figure A3 two parts of the Regulations are shown connected by flow lines to other parts. These are Part 1 on the Scope, Object and Fundamental Requirements for Safety, and Part 2 giving Definitions. An understanding of these parts is essential to the use of the Regulations, and they are dealt with below.

The remaining parts are so interdependent that they are dealt with under their respective headings to discuss the principles, and by means of charts to identify topics which require reference to more than one part.

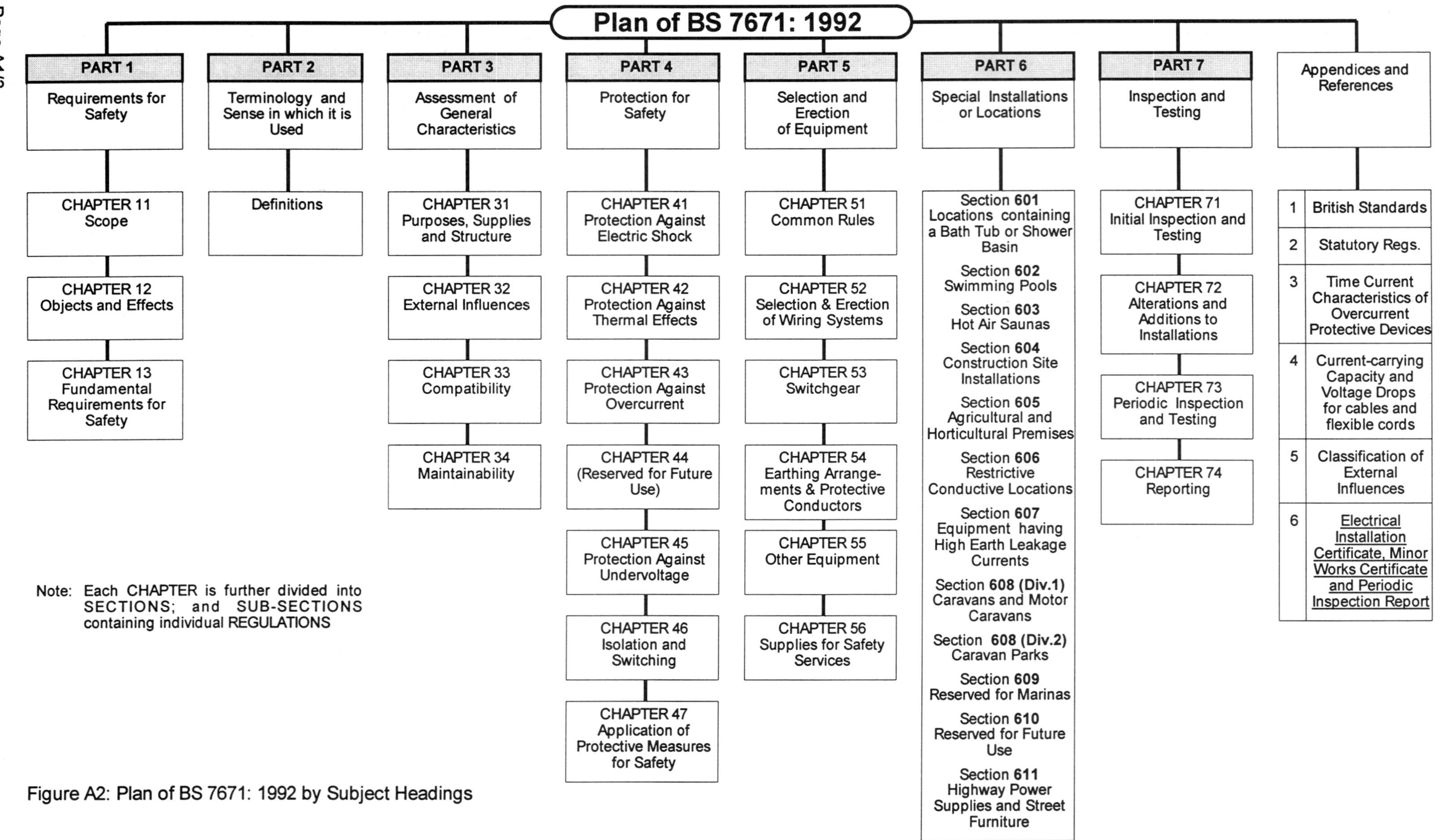

Figure A2: Plan of BS 7671: 1992 by Subject Headings

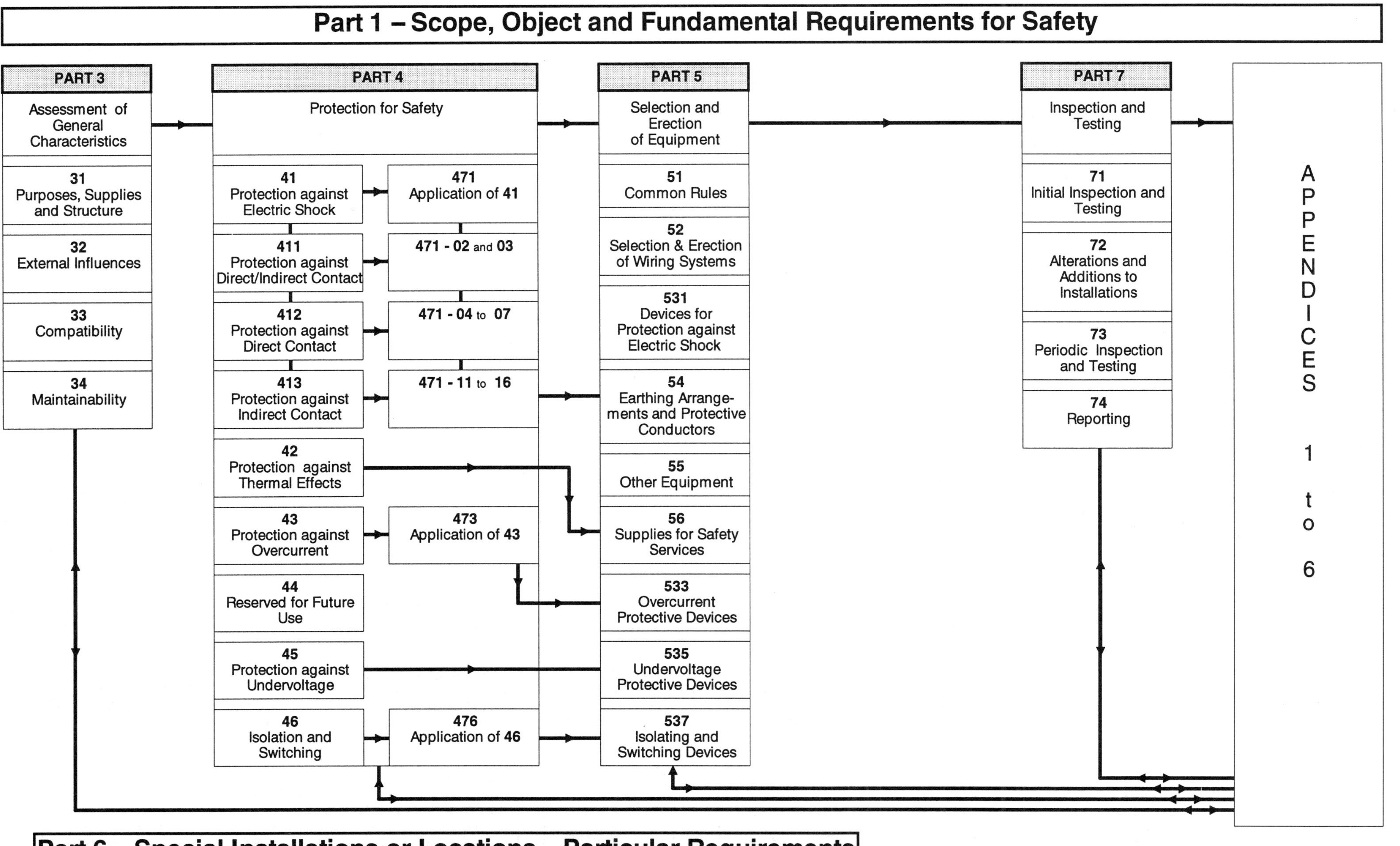

Figure A3: The Plan of BS 7671: 1992 showing Relationship of the Parts of the Regulations

Guide to Interpreting Flow Charts

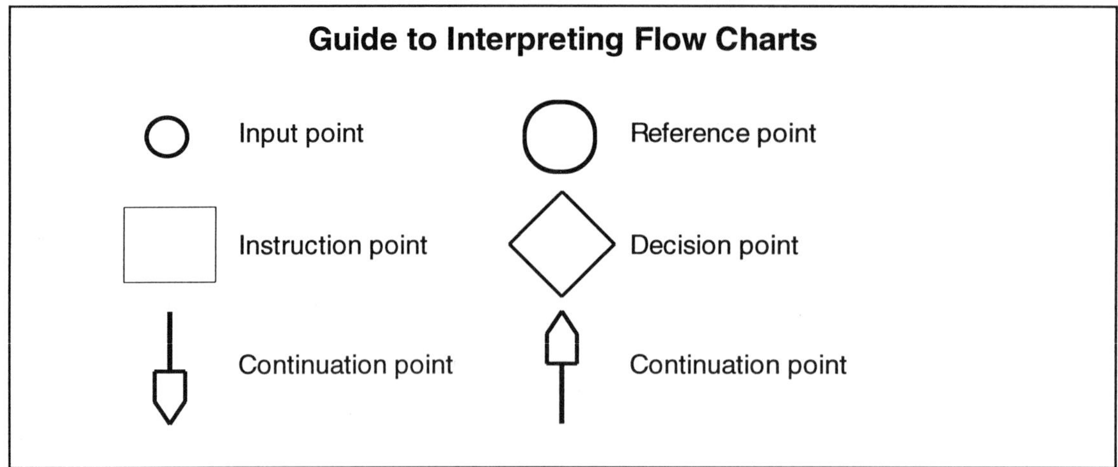

Symbols used Throughout the Handbook

A^2s	=	Amperes-squared per second
C	=	correction factor general
C_a	=	correction factor for ambient temperature
C_d	=	correction factor for type of overcurrent protective device C_d = 1 for HBC fuses and mcbs C_d = 0.725 for semi-enclosed fuses
C_g	=	correction factor for grouping
C_i	=	correction factor for conductors embedded in thermal insulation
C_t	=	correction factor for operating temperature of conductor
I	=	current (general term)
I_a	=	current causing automatic operation of protective device within the time stated
I_b	=	design current of a circuit
I_d	=	fault current of first fault (IT system)
I_n	=	nominal current or current setting of protective device
$I_{\Delta n}$	=	rated residual operating current of the protective device in amperes
I_t	=	tabulated current-carrying capacity of a cable
I_z	=	current-carrying capacity of a cable for continuous service under the particular installation conditions concerned
I^2t	=	energy let-through value of device
I_2	=	current causing effective operation of the overload protective device
k	=	material factor taken from Tables 43A and 54B to F
k^2S^2	=	energy withstand of cable
M	=	appropriate multiplier
R_1	=	resistance of phase conductor from origin of installation to input terminal of the load
R_2	=	resistance of protective conductor from the origin of installation to the earth terminal of the load
S	=	conductor cross-sectional area
t	=	time
Z_e	=	that part of the earth fault loop impedance which is external to the installation
Z_s	=	earth fault loop impedance

N.B. Amendments to the Handbook arising from BS 7671 Amendment No 2 are identified by underlining.

B

The Arrangement of Each Chapter

Scope, Object and Fundamental Requirements for Safety

This PART is divided into three CHAPTERS.

Chapter 11 deals with the scope of application of the Regulations and the types of work excluded.

Chapter 12 deals with the objectives to be achieved through compliance with the Regulations.

Chapter 13 deals with the fundamental requirements for safety encompassing workmanship, materials, conductors, joints, protective devices against overcurrent, earth leakage, and earth fault currents, the positioning of these devices and of the isolation and switching devices required to prevent danger.

Additionally Chapter 13 confirms the necessity for adequate safe access for operation and maintenance, suitability of the whole installation for the environment in which it has to operate, the requirements in regard to additions and alterations, and the need for initial and periodic testing of completed works.

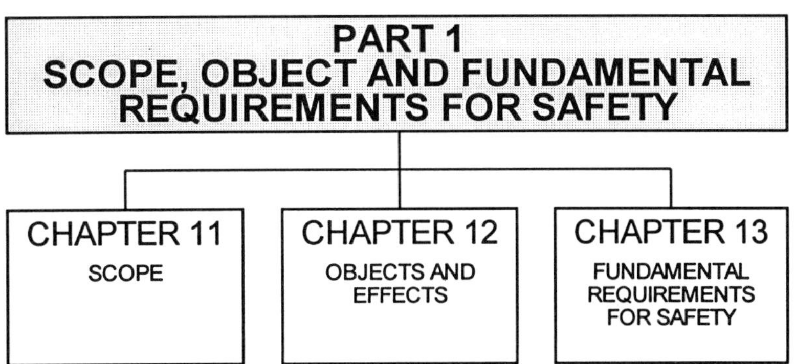

Chapter 11 — Scope

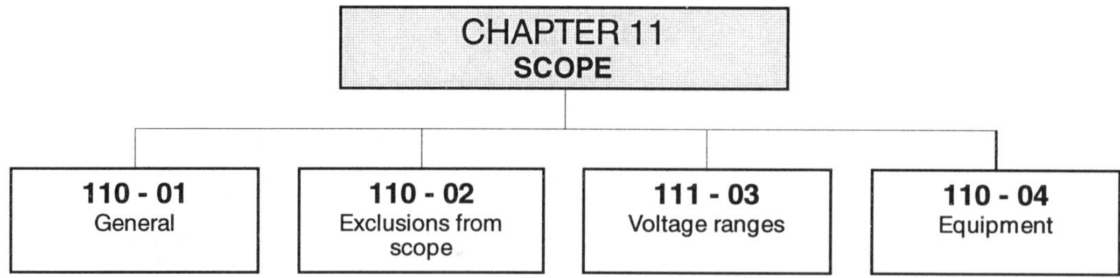

Special Points to Note

(a) The scope does not refer separately to permanent and temporary installations, the Regulations apply equally to both.

(b) The Regulations for the additional and particular requirements for special installations or locations are defined in Section 6.

Chapter 12 — Objects and Effects

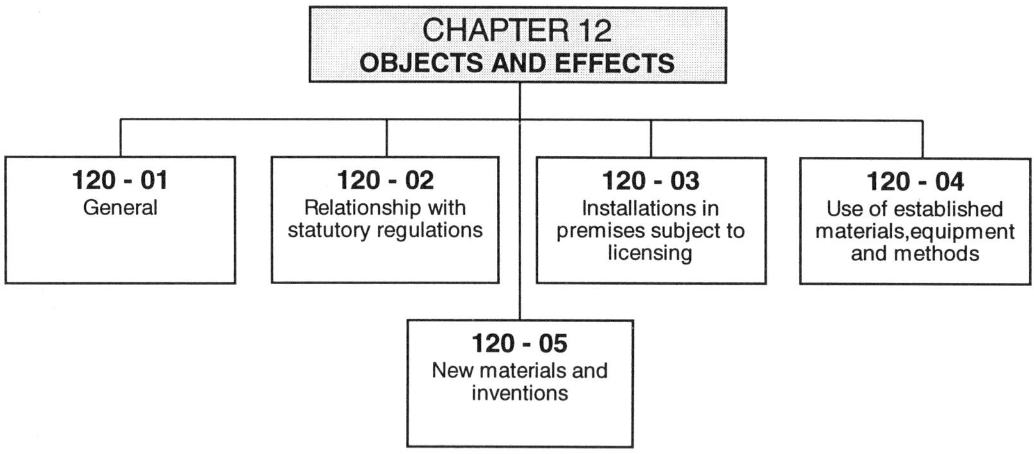

Special Points to Note

(a) The Regulations, although intended to be cited as a whole for contractual purposes, require to be supported by a detailed specification and, on installations of a difficult or special character, the advice of a suitably qualified electrical engineer.

(b) The Regulations are not intended to instruct untrained persons.

(c) Additionally, special requirements, not specifically detailed in the Regulations, apply to installations in premises subject to special licensing or statutory control, e.g. petrol pump/forecourt installations, quarry installations (Appendix 2 lists a number of such situations).

(d) The Regulations include "property" in the protected categories as well as persons and livestock.

(e) The Regulations provide a reminder that there is a need to ascertain and comply with licensing authority requirements.

(f) The Regulations do not include use of new designs that lead to departures.

Chapter 13 — Fundamental Requirements for Safety

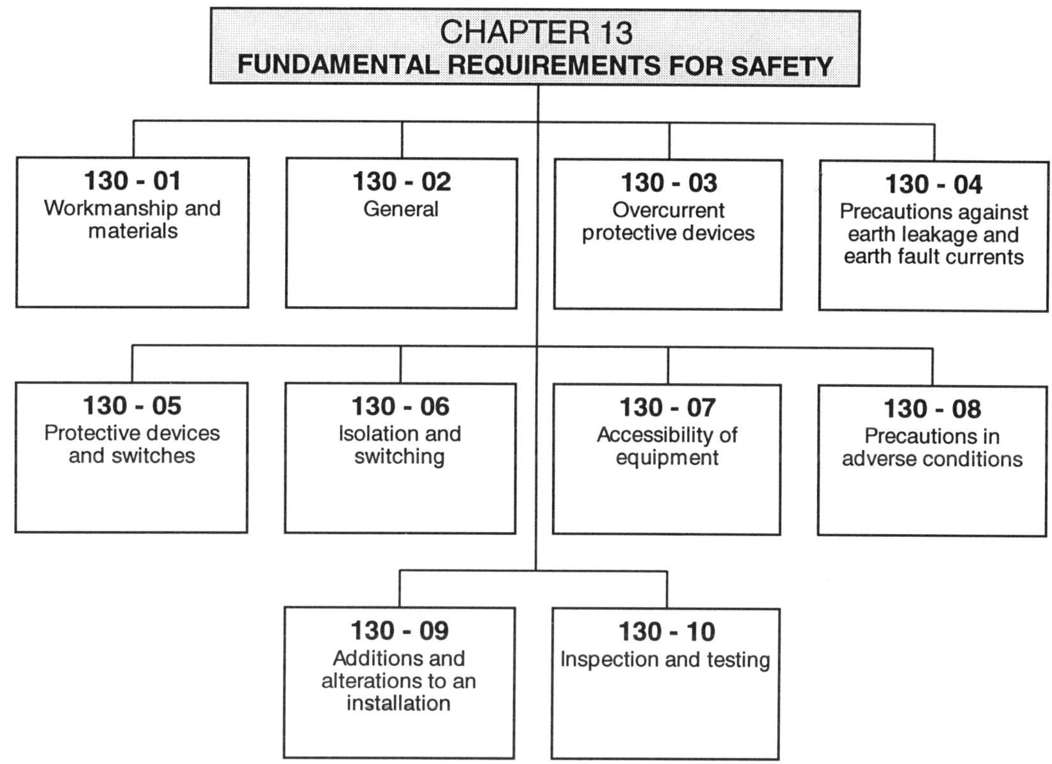

Special Points to Note

(a) The requirement for good workmanship and proper materials is a Regulation (130 - 01 - 01).

(b) It is a requirement (130 - 02 - 01) that future maintenance of the installation be taken into account, and the necessary provisions, such as sub-division for isolation, made.

(c) Regulation 130 - 04 - 02 requires additions and alterations to an installation to comply with all relevant Regulations. Reference must also be made to Regulation 721 - 01 - 01 which imposes further requirements.

(d) Regulation 130 - 10 - 01 emphasises the importance to be accorded to initial and periodic inspection and testing.

(e) Some terms, e.g. overcurrent, residual current device, require reference to the Definitions. It is important that users are familiar with this terminology to avoid any misunderstanding over the meaning and use of the Regulations.

(f) It should be noted that the requirments of Chapter 13 are not only those of the Wiring Regulations, but also in many instances are statutory, mainly under the Electricity Supply Regulations 1988 as amended, and the Electricity at Work Regulations 1989 (see also Appendix 2 of BS 7671).

Definitions

A thorough knowlege of the "Definitions" which form Part 2 of BS 7671: 1992 is essential to ensure common usage and understanding of the terminology in each clause. Care must be taken to ensure accurate usage of the terms by all concerned.

Amendment No. 2 to BS 7671: 1992 – modifies the following definitions:

Building void, accessible
Cable channel
Cable ducting
Cable ladder
Cable trunking
Cable tray
Cable tunnel
Circuit
Distribution circuit
Highway distribution circuit
Reinforced insulation
Symbols used in the Regulations

Amendment No. 2 to BS 7671 – incorporates the following new definitions:

Hazardous-live-part
Ordinary person
Voltage bands

Figure B1: To Supplement the Definition of "System"

Important: The meanings assigned to the various letters in the table below are for explanation only. Other combinations of the various letters to identify special systems is not permissible, unless these have received international approval.

<table>
<tr><th>FIRST LETTER</th><th>SECOND LETTER</th><th>SUBSEQUENT LETTERS (if any)</th></tr>
<tr><td>Earthing arrangement for source of energy</td><td>Relationship of exposed-conductive-parts of installation and earth</td><td>Arrangement of Neutral and Protective Conductors in TN Systems</td></tr>
<tr><td rowspan="3">T
Direct connection of one or more points of the source of energy to earth</td><td rowspan="3">N
Direct connection of exposed-conductive-parts by protective conductors to earthed point of source of energy (usually the neutral point for AC systems)</td><td>C
Neutral and protective conductor combined in a single conductor</td></tr>
<tr><td>S
Separate conductors for neutral and protective functions</td></tr>
<tr><td>C-S
Neutral and protective functions combined in part of system only</td></tr>
<tr><td>T
Direct connection of one point to earth at source of energy</td><td>T
Direct electrical connection of the exposed-conductive-parts by protective conductors to earth via an earth electrode which is independent of the earthing arrangements of the source of energy</td><td></td></tr>
<tr><td>I
All live parts of source of energy isolated from earth or one point earthed through an impedance</td><td>T
Direct electrical connection of the exposed-conductive-parts by protective conductors to earth via an earth electrode which is independent of the earthing arrangements of the source of energy</td><td></td></tr>
</table>

NOTES (i) An example of the use of a combined neutral and protective conductor (PEN Conductor) is earthed concentric wiring but special authorisation must be obtained from the appropriate authority.

(ii) Special requirements apply to installations supplied from PME systems (TN-C, TN-C-S) and the Supplier concerned should be consulted.

(iii) The 'Electricity Supply Regulations 1988 as amended', do not permit the use of IT systems for public supply.

(iv) Different systems may be encountered in the same building or location. For example, there may be extra low voltage circuits for a particular function and because of the nature of the source of supply (e.g. batteries) this will be a separate system from the general building system.

Assessment of General Characteristics

PART 3 is concerned with the aspects and purpose of every installation, and must be considered in detail before any design or installation is started.

The five chapters of PART 3 should not be regarded merely as a check-list, but rather as a step-by-step guide, each item of which must be examined carefully.

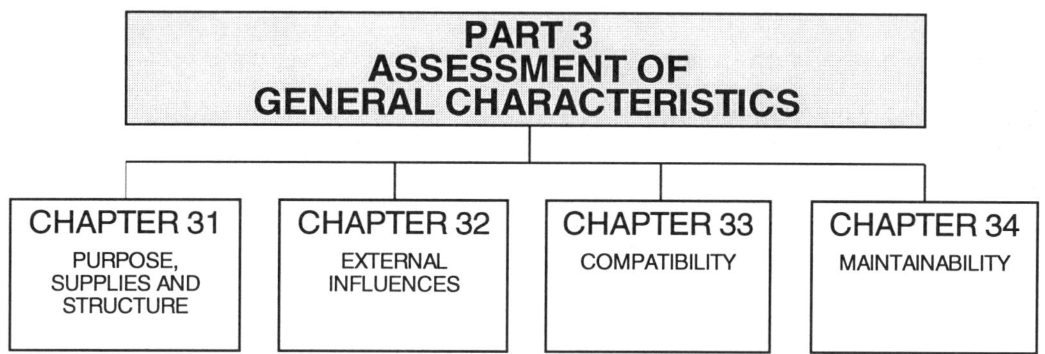

Special Point to Note

External influences – See also Regulations 512 - 06 - 01 and 02

Compatibility – Regulation 331-01-01 includes a list of some of the characteristics likely to have harmful effects.

Chapter 31 — Purpose, Supplies and Structure

Study of this chapter will provide the governing parameters for assessing loads, the system (see Part 2), number and type of conductors and the requirements for standby supplies, together with earthing.

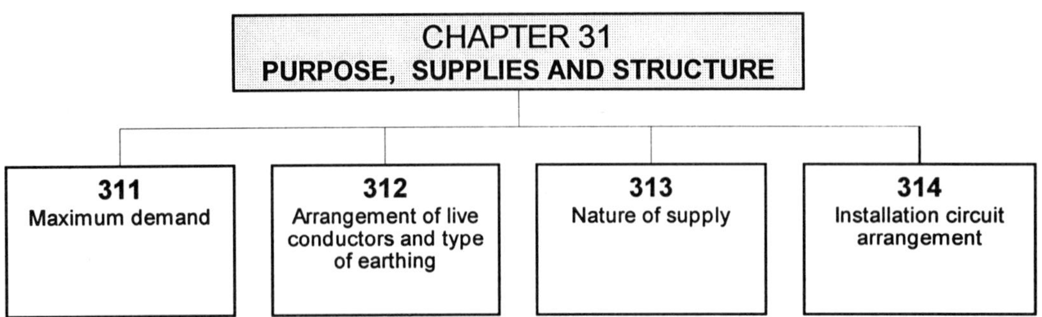

PART 4

OF THE REGULATIONS

Protection for Safety

Having considered the purpose for which the installation is being designed (PART 3), the designer and installer must concern themselves with the safety of the installation from the point of view of protecting the users, the property and the installation.

The hazards to be guarded against are electric shock, fire, burns, and the effects on the installation of overload current, fault current and undervoltage. The necessary measures must also be taken to ensure that the installation can be controlled properly and operated safely.

Chapters 41, 42, 43, 45 and 46 are concerned with the basic measures to be adopted for protection and Chapter 47 deals with their application. Reference is also necessary to PART 5 – Selection and Erection of Equipment – and to PART 7 – Inspection and Testing.

Whilst familiarity with PART 4 is essential, the presentation of the various protective measures as Topic Charts and Illustrations, later in this Handbook, is intended to draw together the various requirements for ease of reference.

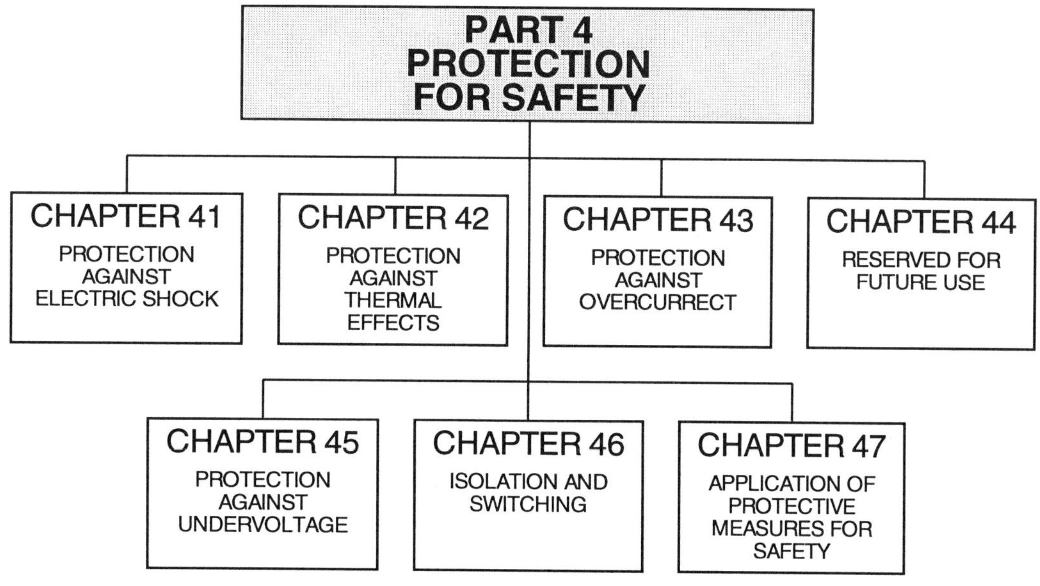

Chapter 41 — Protection against Electric Shock

This chapter is concerned with protection of the users of the installation, whether they be concerned with its operation or availing themselves of its facilities. The provisions are qualified and to some extent explained by Chapter 47 and reference must be made to this chapter.

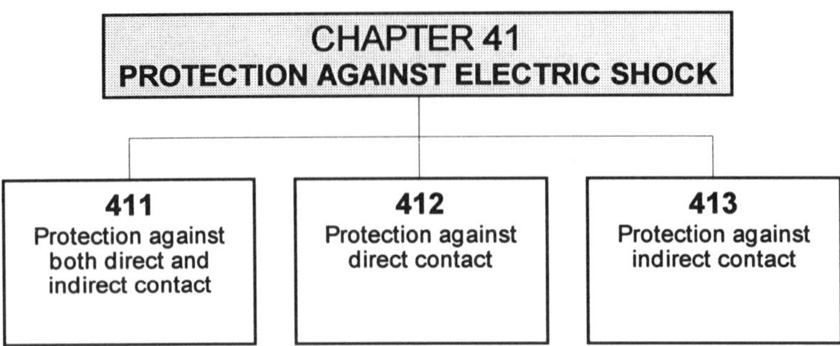

Special Points to Note

(a) The requirement in Regulation 413 - 02 - 05 to allow for the change in conductor resistance due to temperature rise is deemed to be satisfied if the protective device complies with Appendix 3 and the circuit loop impedance is within the appropriate values in Regulation 413 - 02.

(b) The range of Z_S values and cpc impedances are derived from BS 3871 and BS EN 60898: 1991.

(c) Loop impedance values in tables are related to normal operating temperatures. If the conductors are at a different temperature when tested an appropriate adjustment must be made.

(d) When using Table 41E only phase to phase voltage is available if the neutral is not distributed, i.e. connected to the load.

(e) 412-02-01 refers to stresses to which insulation may be subjected for non-factory built equipment.

(f) 413-02-04 provides details of the protection measures to be used if the conditions for automatic disconnection cannot be met by an overcurrent device.

Chapter 42 — Protection against Thermal Effects

This chapter is concerned with the measures to be adopted to prevent danger from fire and burns which may be caused by the heat generated by fixed electrical equipment.

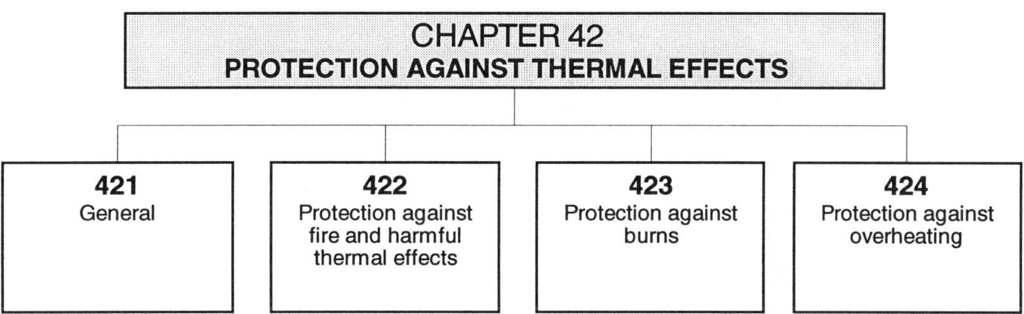

Special Points to Note

(a) This chapter places an onus on the designer to decide on safe distances or materials to provide protection against thermal effects.

(b) Regulation 422 - 01 - 04 limits the use of open back consumer units, in addition see Section 526 - 03.

(c) Regulation 423 - 01 - 01 recognises that the same material in different situations can be allowed to reach different temperatures before being dangerous.

(d) Regulation 423 - 01 - 01 states that certain fixed appliances, e.g. storage heaters, cannot be regarded as being sufficiently guarded to prevent accidental contact. Nevertheless, having regard to their purpose and if they comply with the appropriate British Standard, their use is admissible.

(e) A requirement to ensure air flow is present when the heating element of a forced air heater is on, together with two temperature limiting devices is inherent in Regulation 424 - 01 - 01.

Chapter 43 — Protection against Overcurrent

This chapter is mainly concerned with the factors which must be considered at the design stage of an installation. The chapter must be read in conjunction with other chapters (see Figure A3) and Topic Chart 5 – Overcurrent.

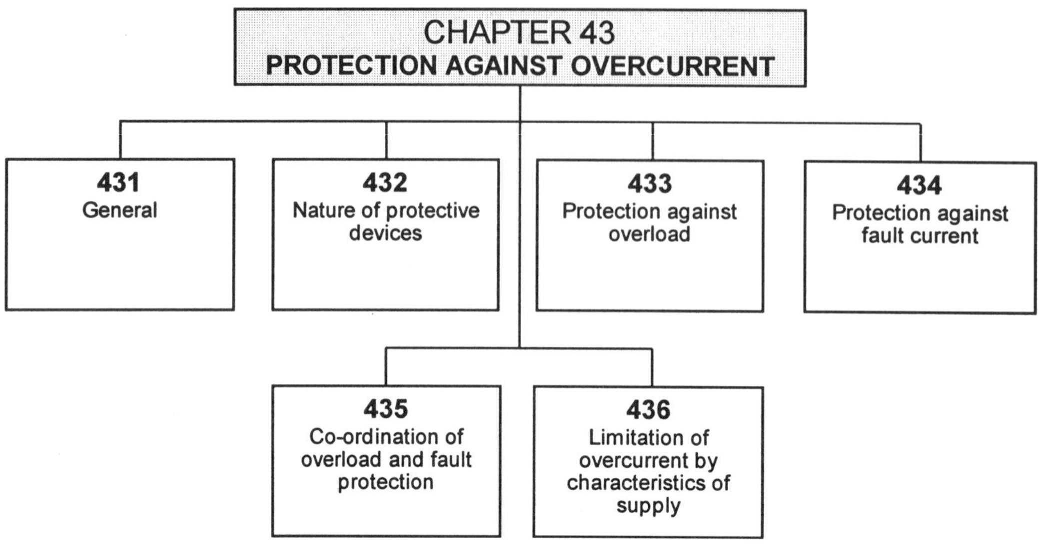

Special Points to Note

(a) Regulation 433 - 02 - 04 sets out the deemed to comply with Regulation 433 - 02 - 01 requirements, for ring final circuits protected by a 30A or 32A protective device.

(b) Regulation 434 - 04 - 01 requires a calculation to be made to check the short circuit withstand of conductors and cpc for a fault not affecting all parallel conductors thus possibly extending the disconnection time.

Chapter 45 — Protection against Undervoltage

This chapter is concerned with the prevention of possible danger and/or damage caused by supply voltage reduction, or cessation and subsequent resumption of supply. It should be noted that Regulation 552 - 01 - 03 is also relevant.

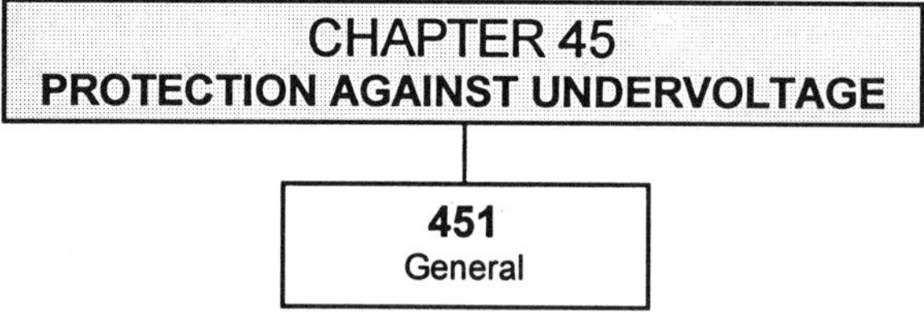

Chapter 46 — Isolation and Switching

This chapter is concerned with the basic requirements to prevent or remove hazards associated with the electrical installation or electrical equipment and machines.

The chapter must be read in conjunction with other chapters (see Figure A3) and Topic Chart 4 Isolation and Switching.

CHAPTER 46
ISOLATION AND SWITCHING

460	461	462	463	464
General	Isolation	Switching off for mechanical maintenance	Emergency switching	Functional switching (control)

Sections 461, 462 and 463

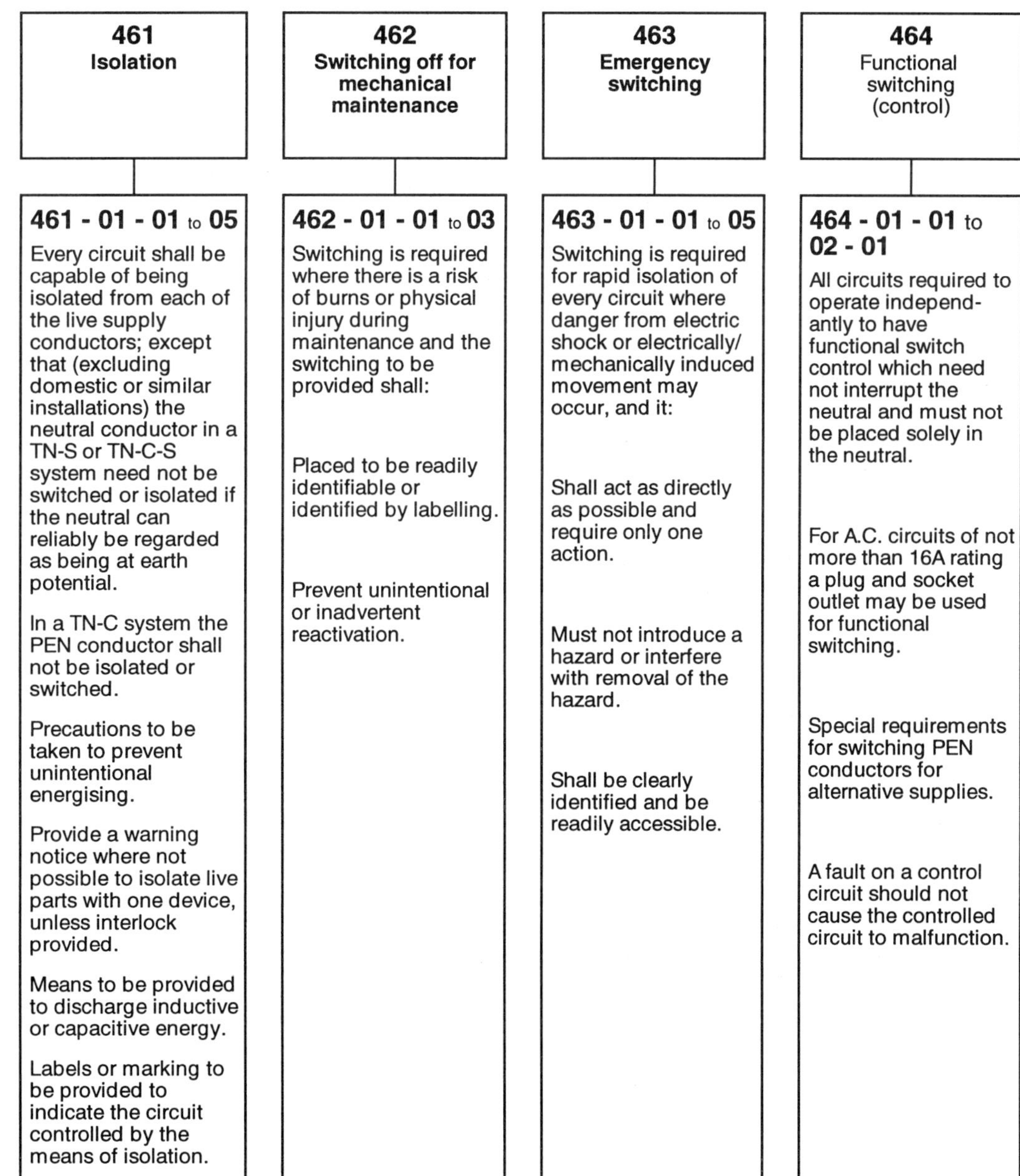

Special Points to Note

(a) The protective or PEN conductor shall not incorporate any means of isolation or switching (460 - 01 - 03) except as 460 - 01 - 05 and in a TN-C system shall not be isolated or switched.

(b) Provision need not be made for isolation of the neutral conductor in TN-S and TN-C-S systems unless specifically required by 460 - 01 - 02.

Chapter 47 — Application of Protective Measures for Safety

This chapter is extremely important for it is here that the qualification and, in some cases, amplification of the requirements of Chapters 41, 43 and 46 are contained.

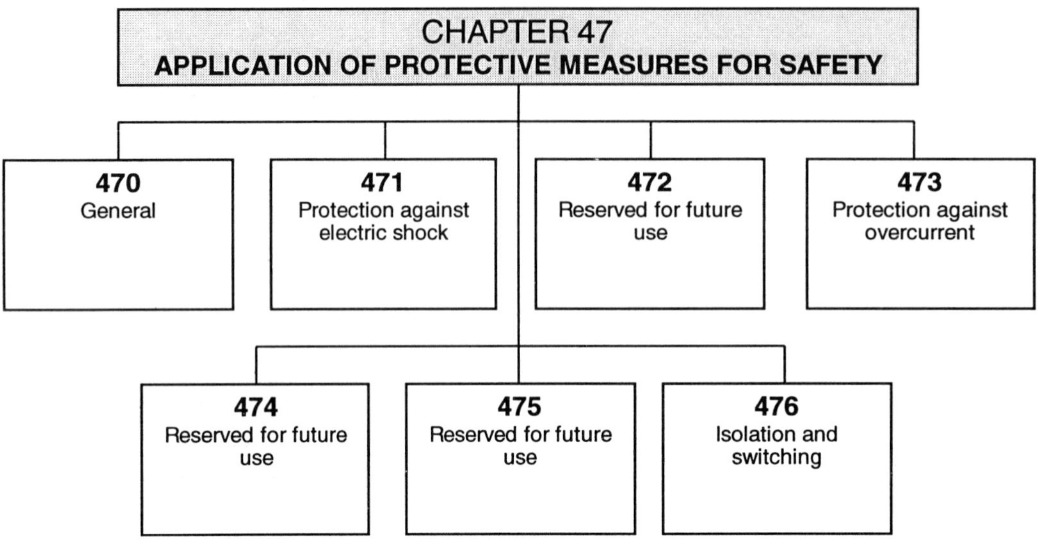

Special Point to Note

No reference is made in Chapter 47 to undervoltage protection, which is dealt with in Chapter 45 and Regulation 552 - 01 - 03.

Section 471

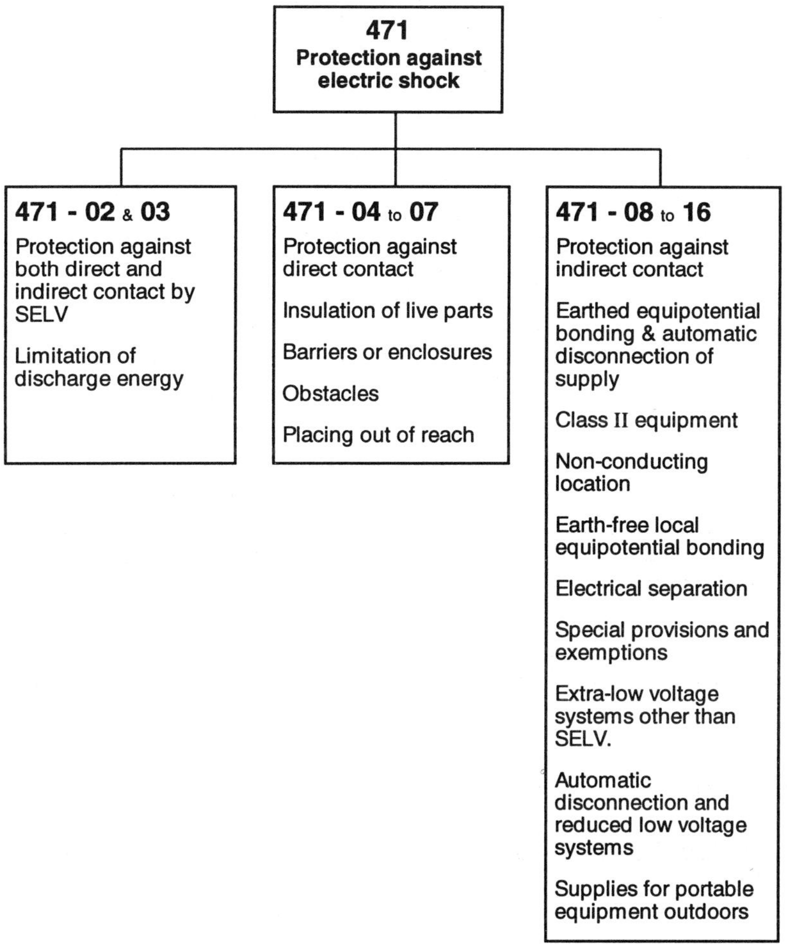

Section 473

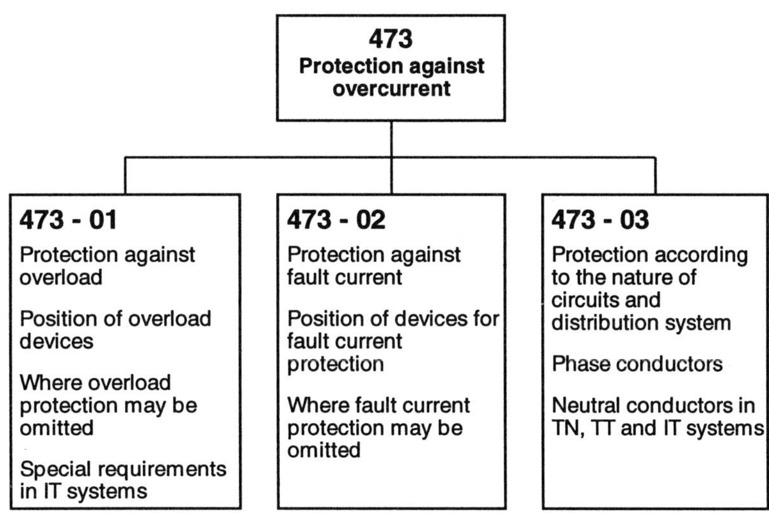

Special Points to Note

(a) Regulation 473 - 01 - 04 now allows the Supplier's device to provide overload protection where the agreement of the Supplier is obtained.

(b) Regulation 473 - 03 - 02 covers special requirements for TT systems.

Section 476

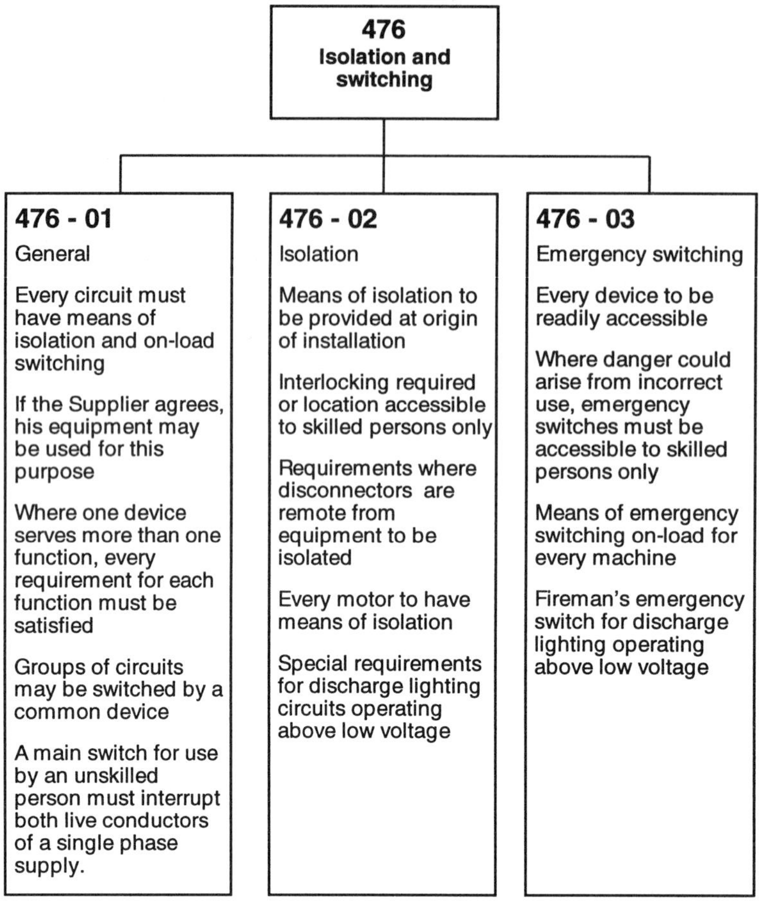

476
Isolation and switching

476 - 01
General

Every circuit must have means of isolation and on-load switching

If the Supplier agrees, his equipment may be used for this purpose

Where one device serves more than one function, every requirement for each function must be satisfied

Groups of circuits may be switched by a common device

A main switch for use by an unskilled person must interrupt both live conductors of a single phase supply.

476 - 02
Isolation

Means of isolation to be provided at origin of installation

Interlocking required or location accessible to skilled persons only

Requirements where disconnectors are remote from equipment to be isolated

Every motor to have means of isolation

Special requirements for discharge lighting circuits operating above low voltage

476 - 03
Emergency switching

Every device to be readily accessible

Where danger could arise from incorrect use, emergency switches must be accessible to skilled persons only

Means of emergency switching on-load for every machine

Fireman's emergency switch for discharge lighting operating above low voltage

Selection and Erection of Equipment

This PART is concerned with the materials and methods of installation which are necessary to attain the standard of safety required by the Regulations.

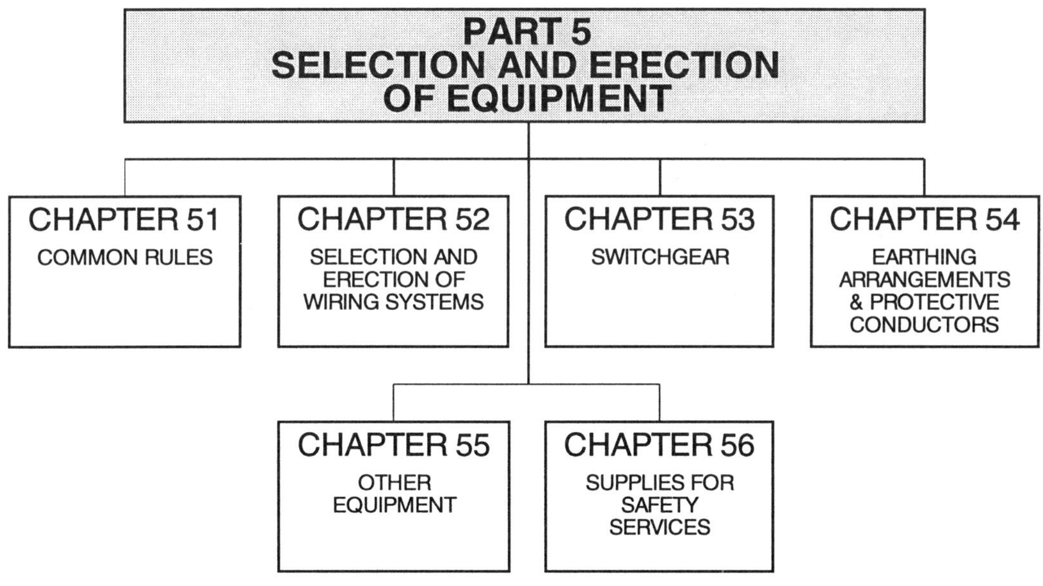

Chapter 51 — Common Rules

In this chapter the rules that apply to all electrical installations, for any purpose or location, are set down as the basic requirements.

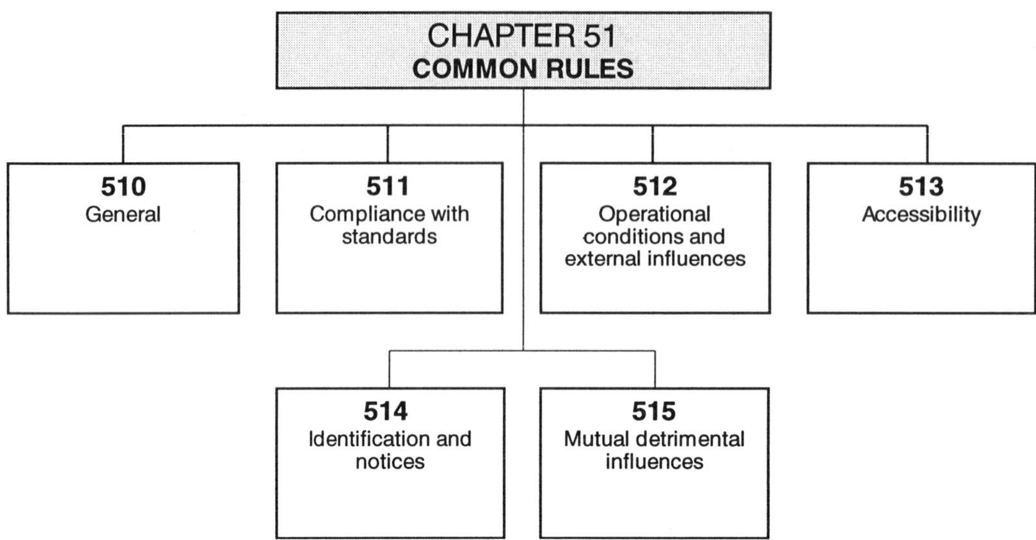

Section 512

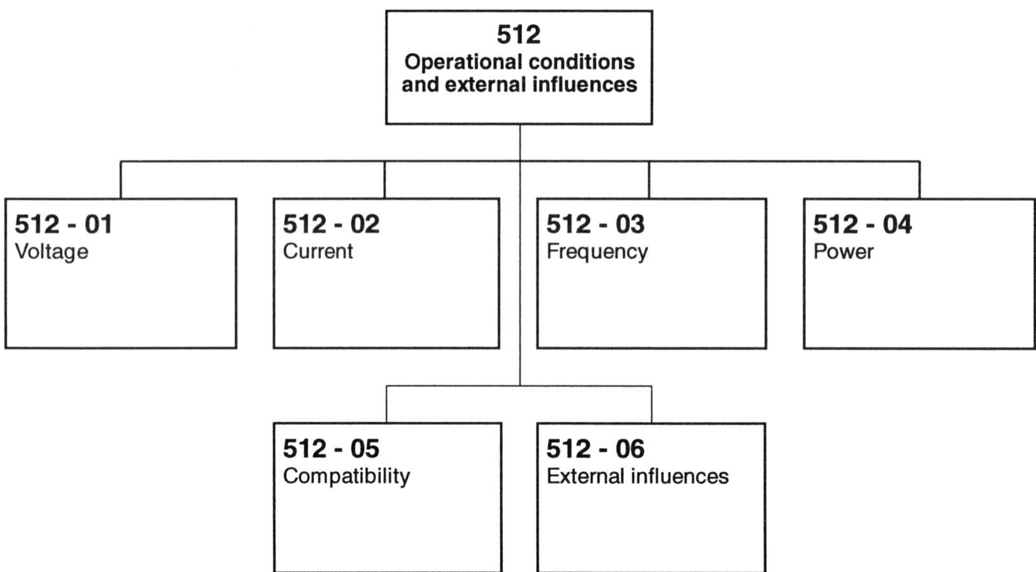

Section 514

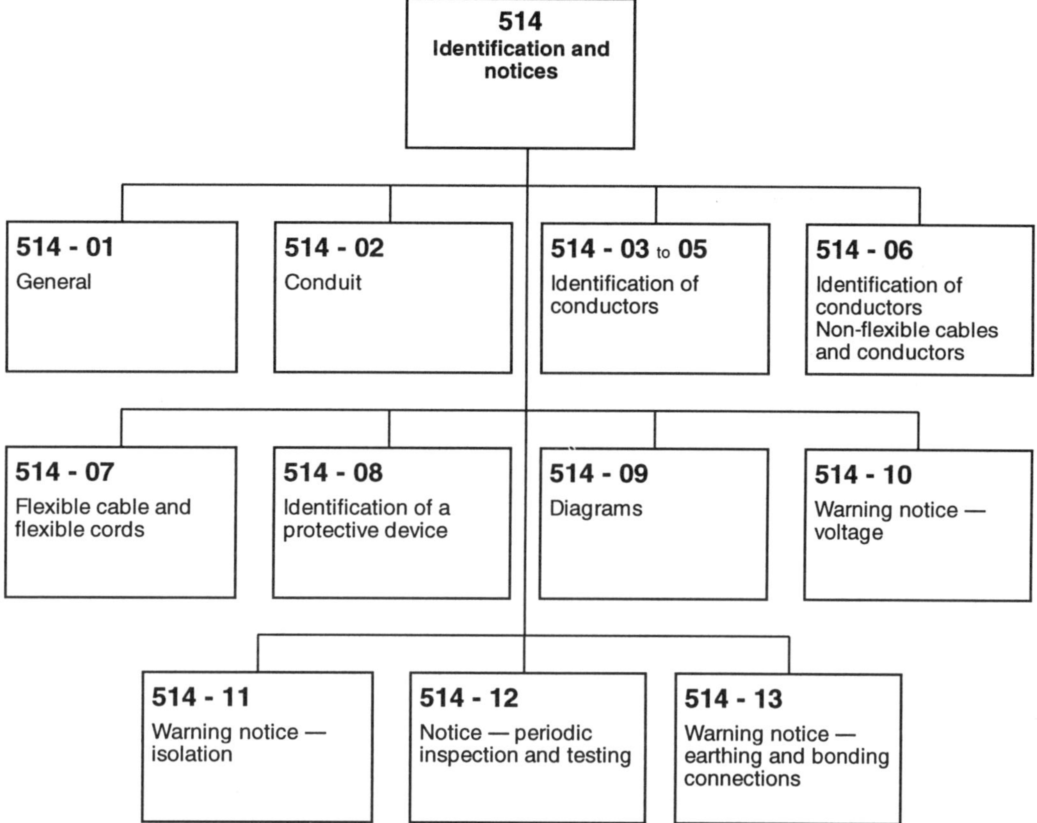

Special Point to Note:

Regulation 514 - 01 - 02 requires identification of wiring by arrangement or marking for inspection

Section 515

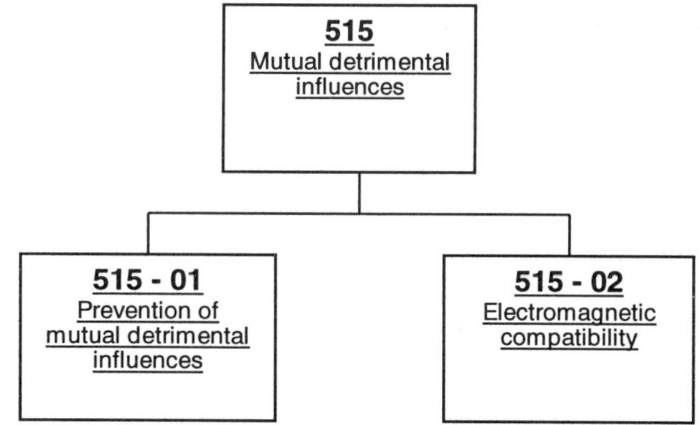

Chapter 52 — Selection and Erection of Wiring Systems

All installations need a wiring system and this chapter contains the requirements that govern the type of wiring system to be selected.

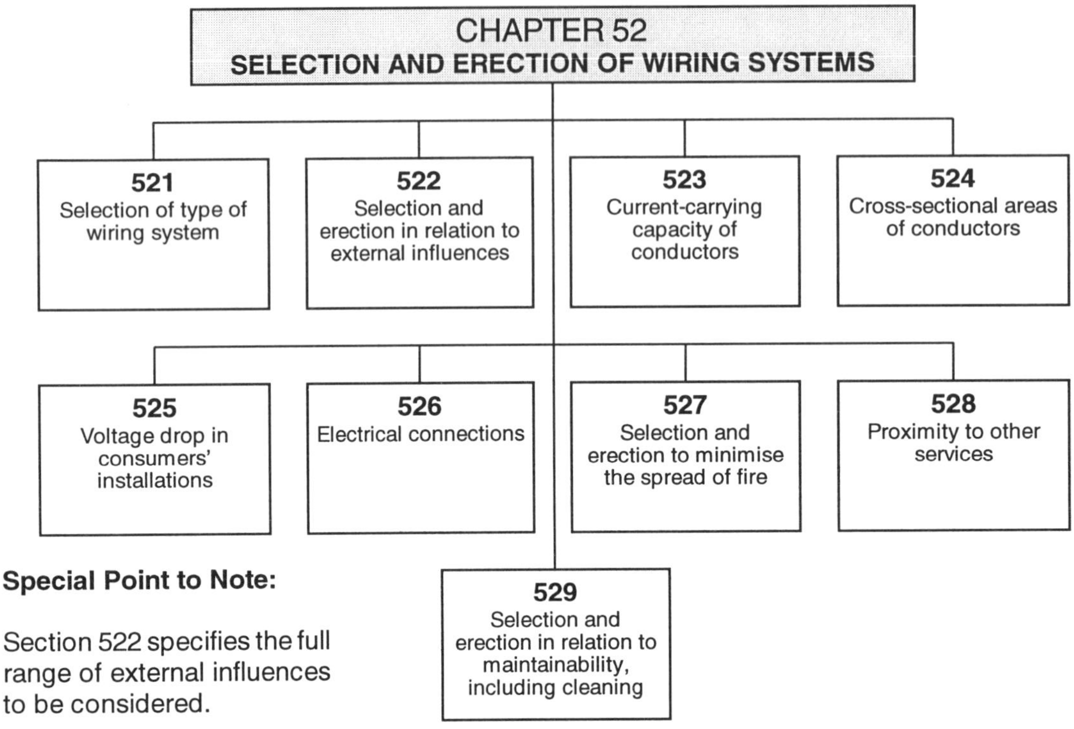

Special Point to Note:

Section 522 specifies the full range of external influences to be considered.

Section 521

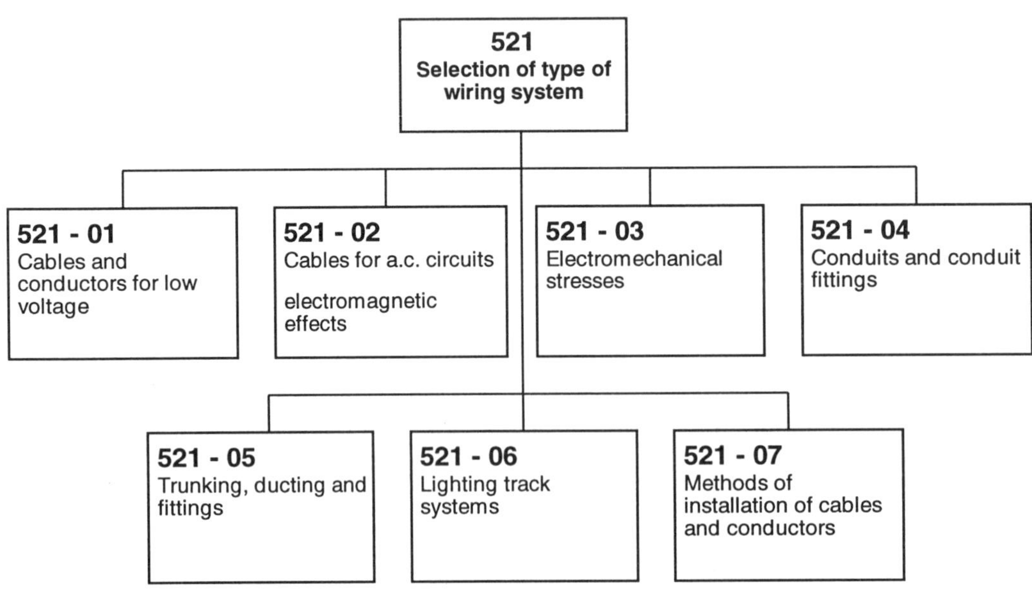

Section 522

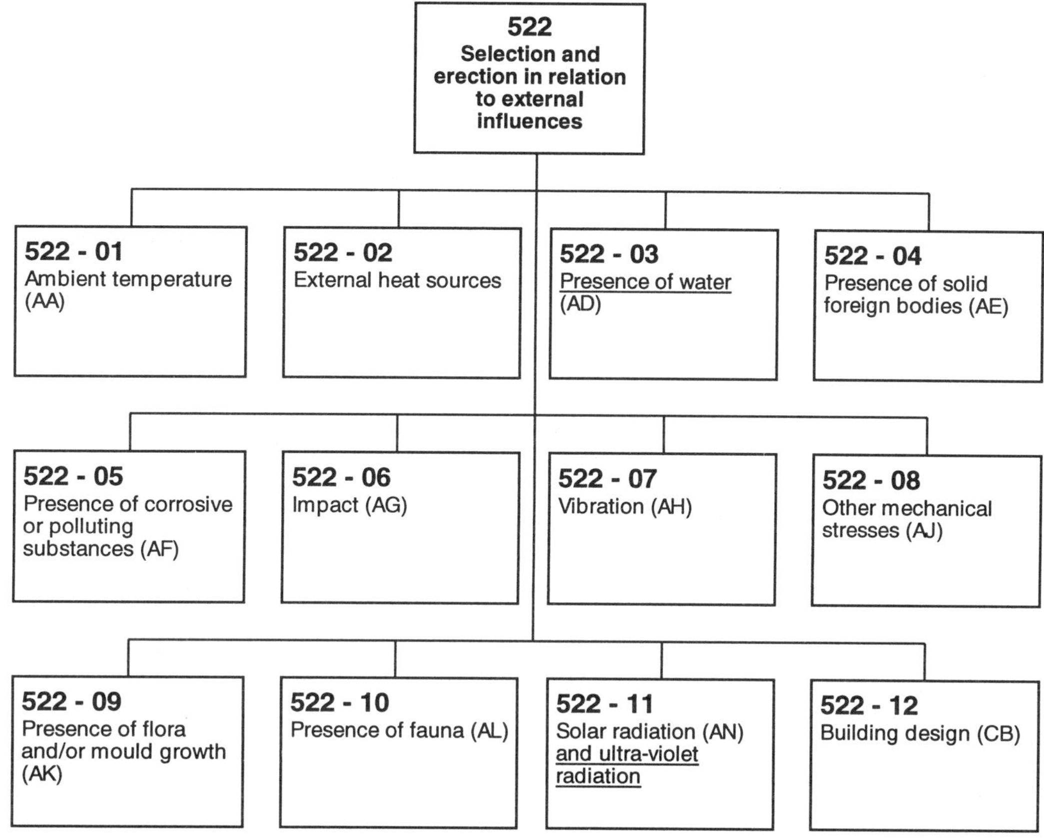

Section 523

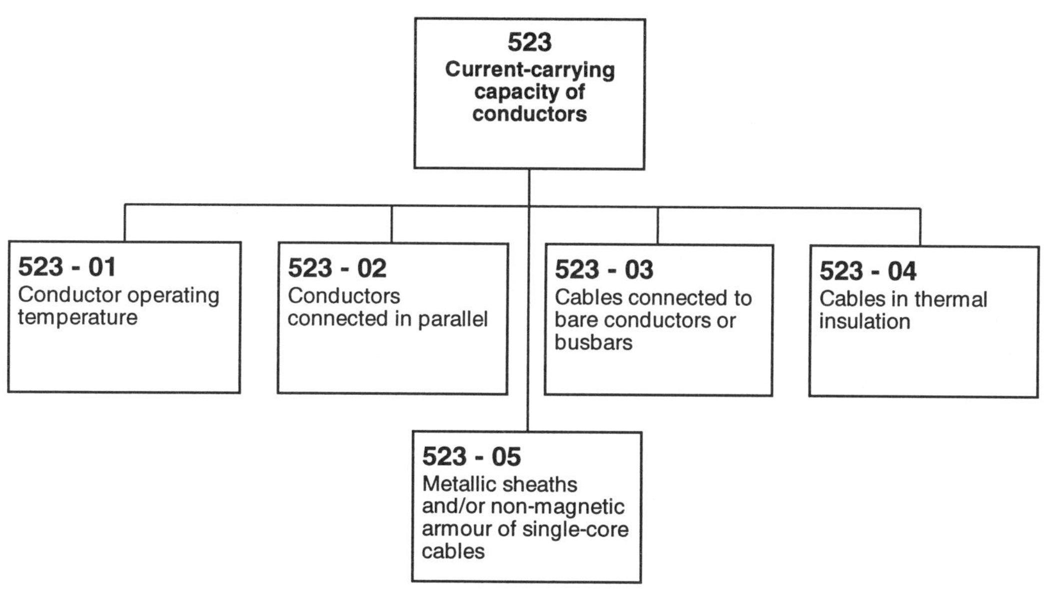

Section 524

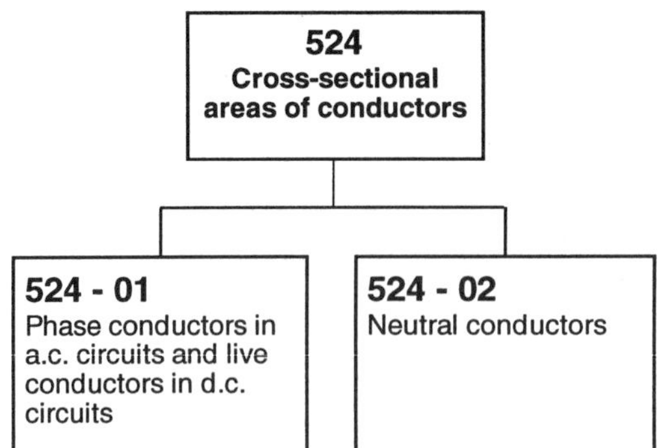

Section 526

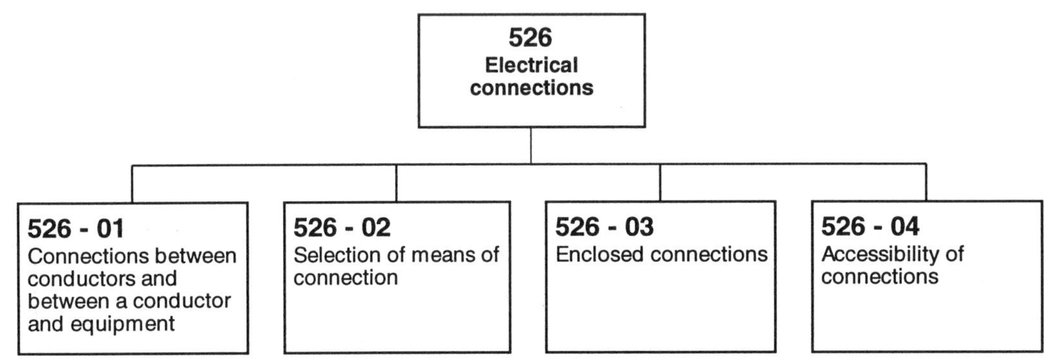

Section 527

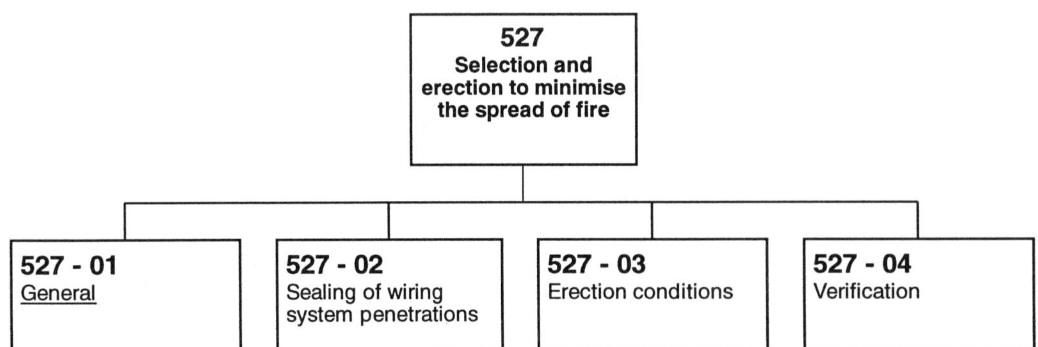

Section 528

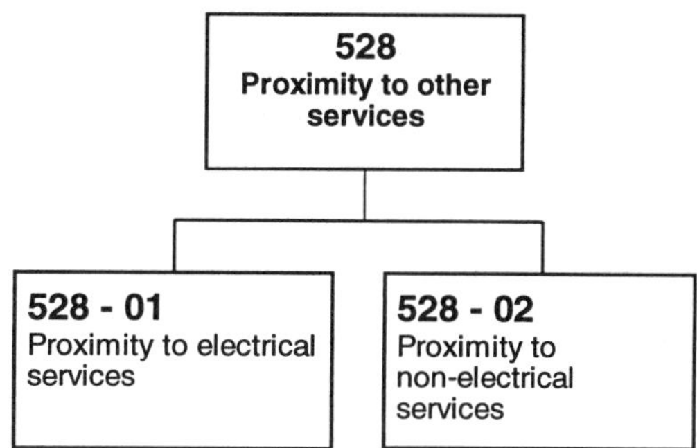

Chapter 53 — Switchgear

Before a selection of switchgear can be made, many factors have to be assessed for their influence on the choice. These factors are set out in this chapter.

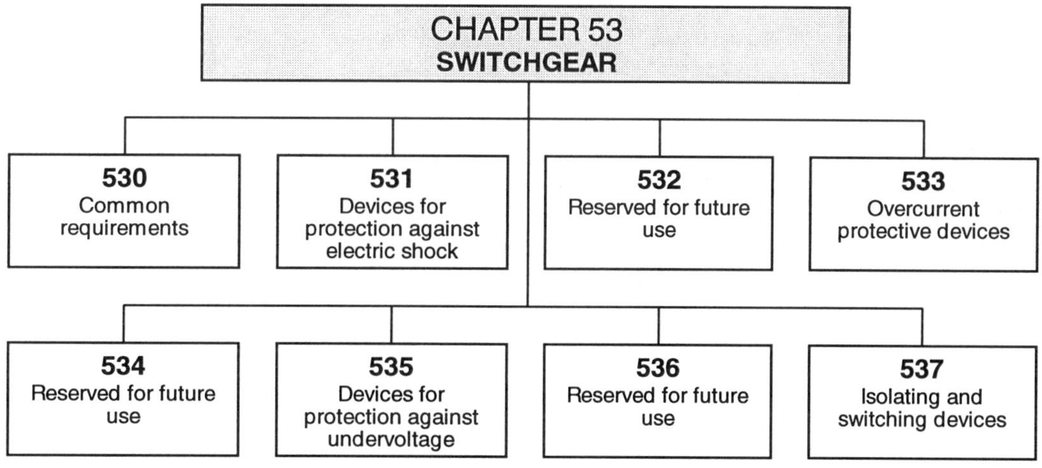

Section 531

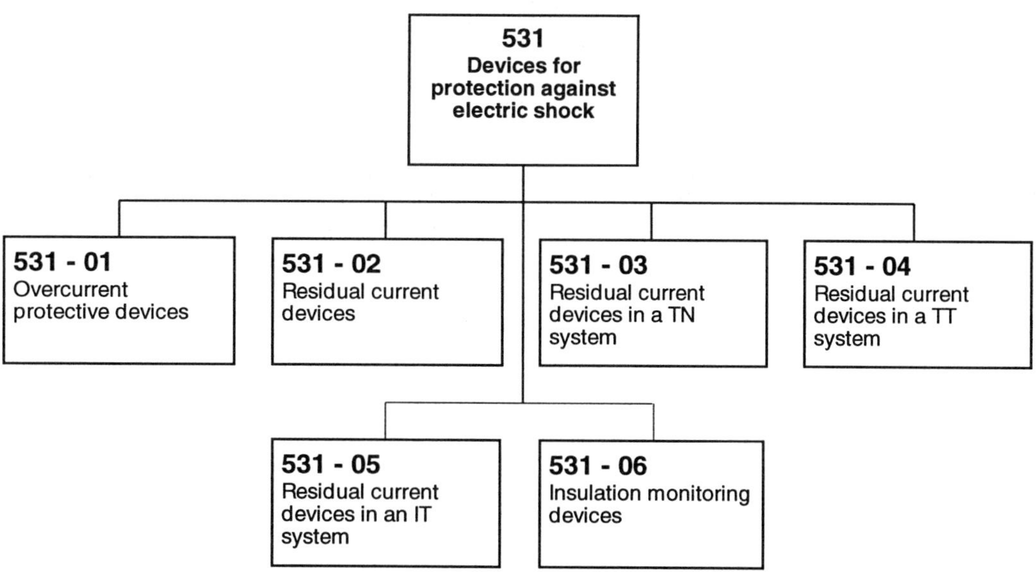

Section 533

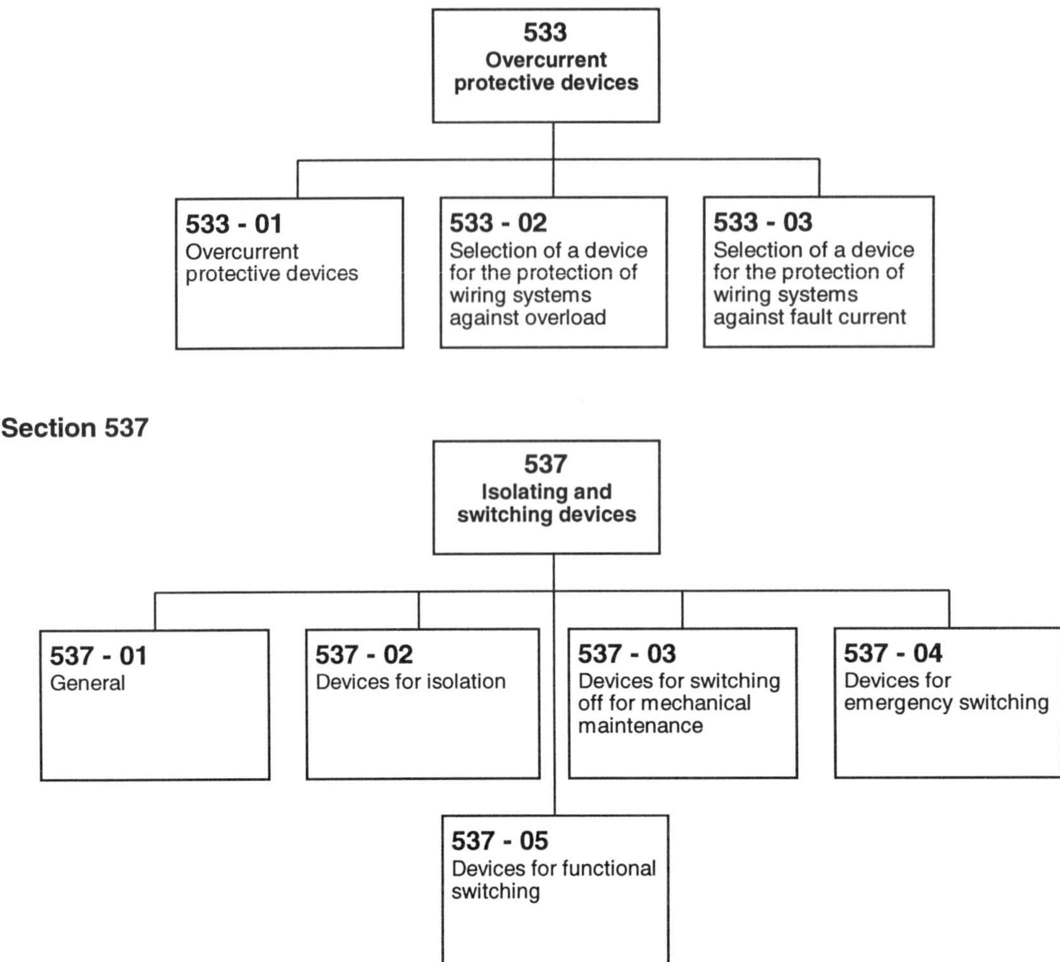

Section 537

Chapter 54 — Earthing Arrangements and Protective Conductors

In order to ensure that protective devices operate, specific earthing arrangements are required supplemented by protective conductors. The applicable methods can be determined by the application of the Regulations in this chapter.

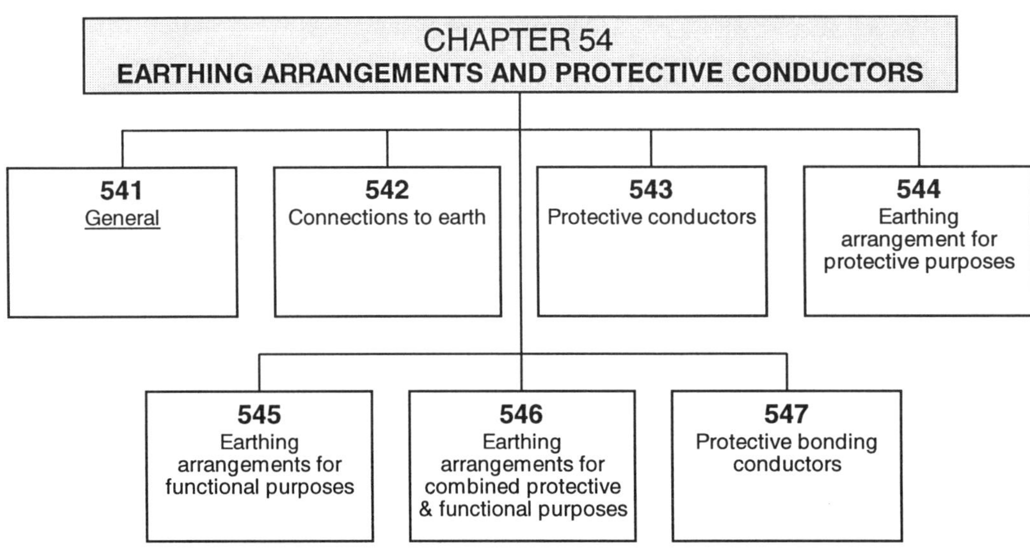

Special Point to Note

(a) If a separate protective conductor is used it shall be installed in close proximity to the cable containing the live conductors (544 - 01 - 01)

Section 542

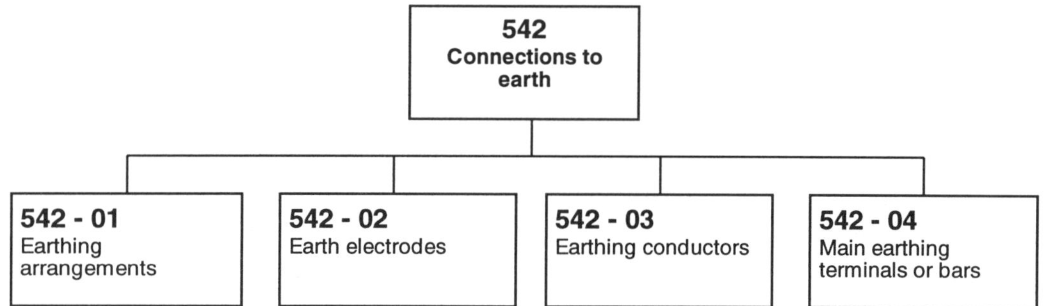

Section 543

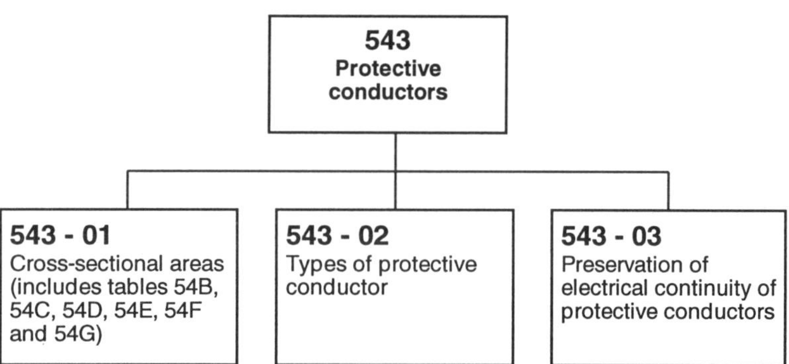

Section 547

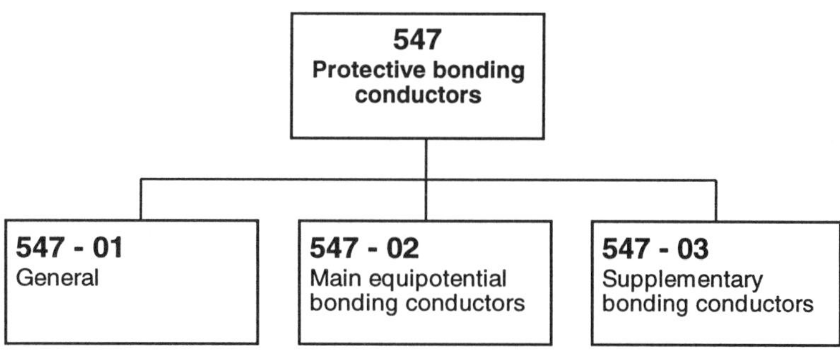

Chapter 55 — Other Equipment

All items of equipment connected to an installation have to meet the requirements set out in this chapter.

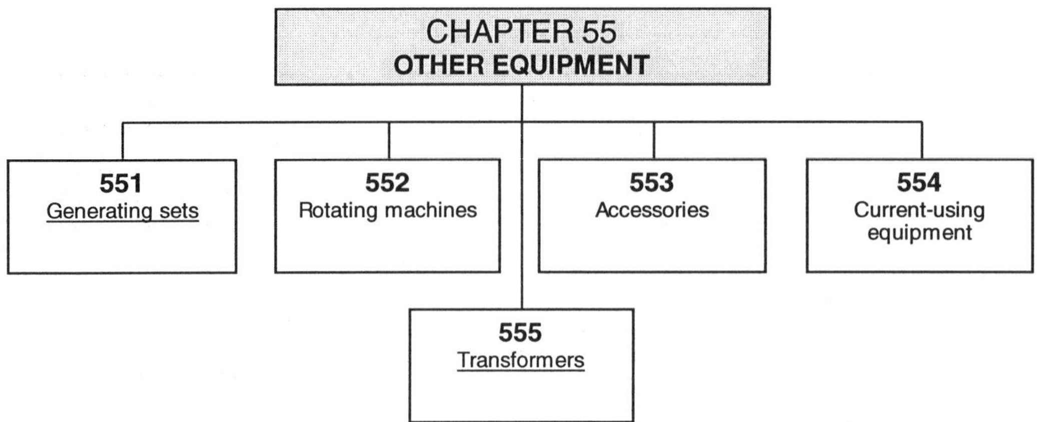

Section 551

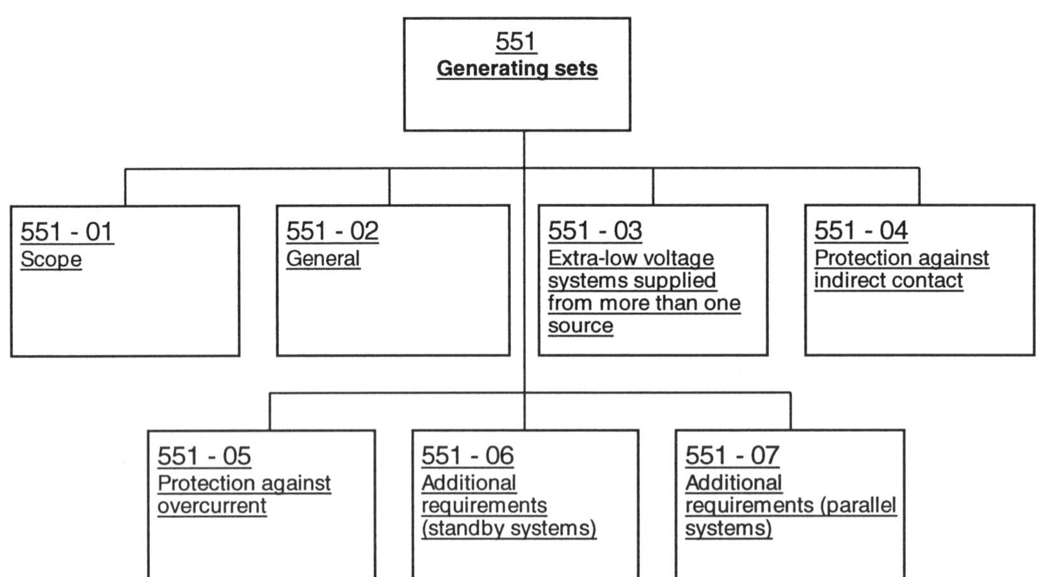

Section 553

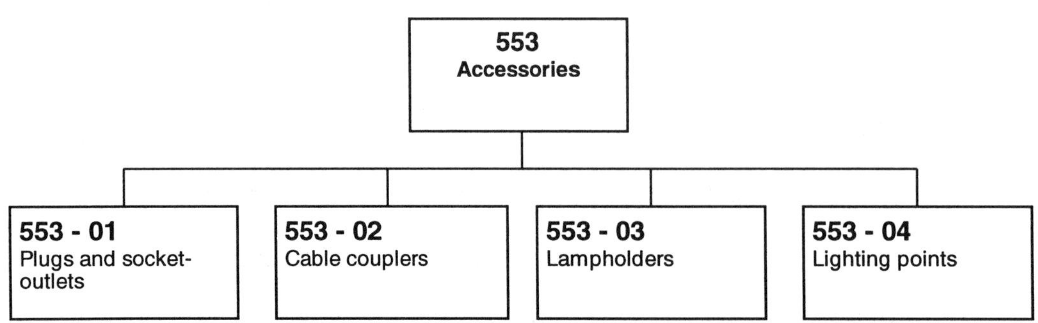

Section 554

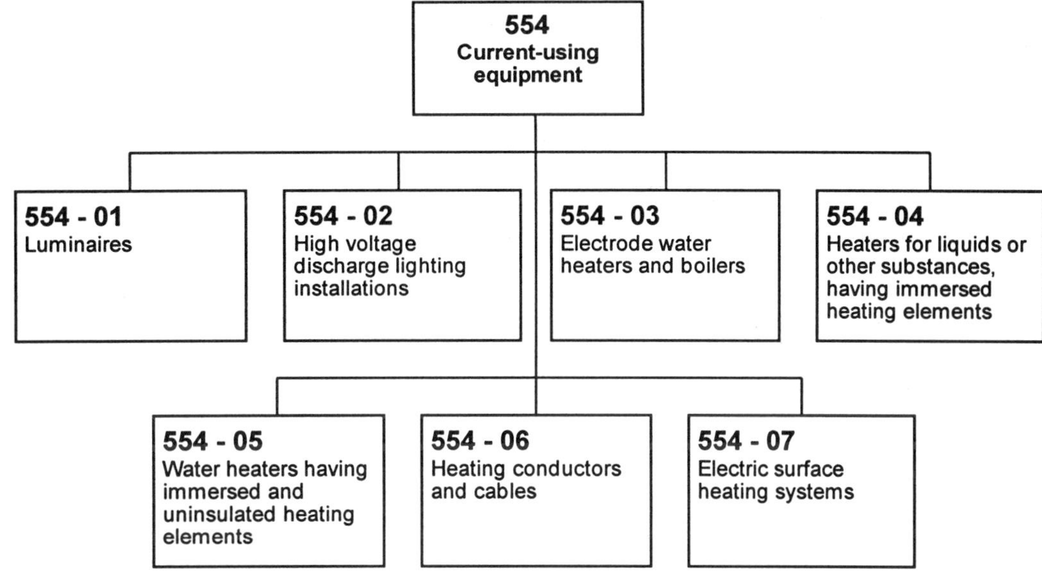

Chapter 56 — Supplies for Safety Services

Within most non-domestic buildings there are circuits that must continue to operate in an emergency such as fire detection and alarm, smoke ventilation fans, sprinkler pumps, etc. The supplies to these safety services need to comply with the special requirements of this chapter to ensure they remain working for as long as possible.

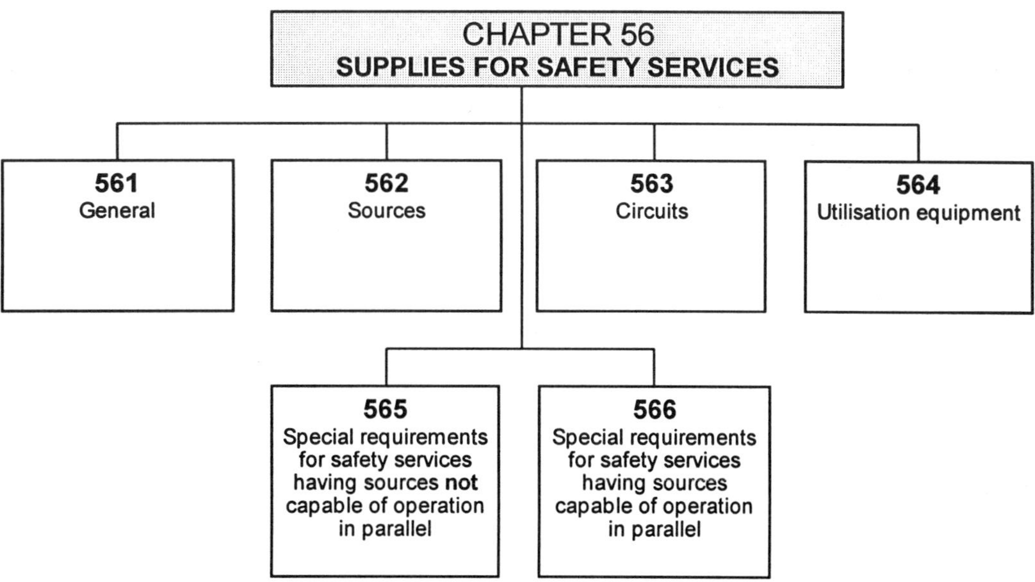

Special Points to Note

(a) Equipment to be located to facilitate inspection, testing and maintenance.

(b) Exhaust fumes etc, from a source must not cause a hazard.

(c) Regulation 566 - 01 - 02 draws attention to circulating currents caused by harmonics.

PART 6

OF THE REGULATIONS

Special Installations or Locations – Particular Requirements

Parts 1 - 5 give the designer or installer all the requirements for installations in standard locations either Commercial, Industrial or Domestic.

However, there are special locations either within a standard installation or existing in their own right. These special installations or locations by their nature or the nature of the activity that takes place within them, need special attention.

This part of the Regulations supplements or modifies the requirements of other sections of the Regulations.

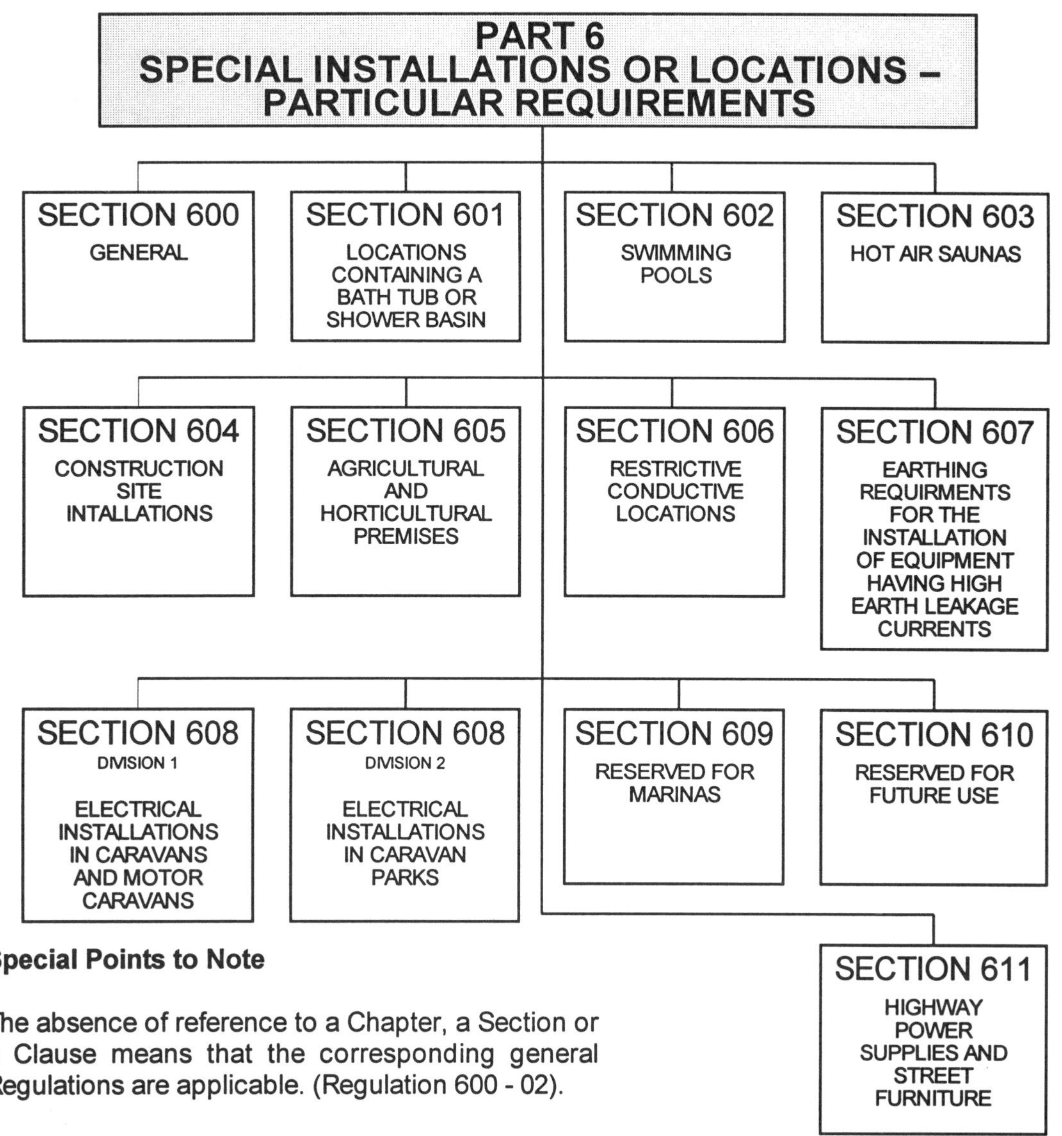

Special Points to Note

The absence of reference to a Chapter, a Section or a Clause means that the corresponding general Regulations are applicable. (Regulation 600 - 02).

Section 601

A room containing a bath tub or shower basin is there for the express purpose of allowing people to be wet, thereby reducing their body resistance. To minimise the possibility of electric shock under these conditions, the requirements of this chapter must be incorporated in the installation.

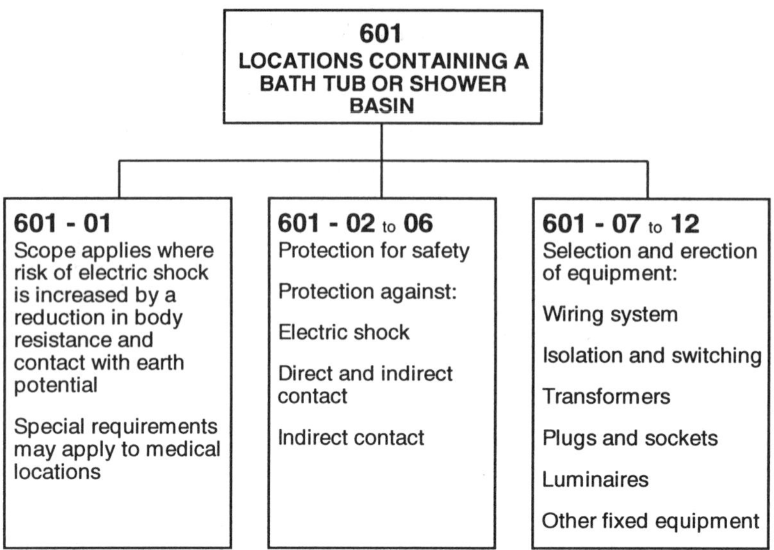

Section 602

The comments of Section 601 apply to this chapter with the additional risk of persons using electrical equipment on the poolside or in the water.

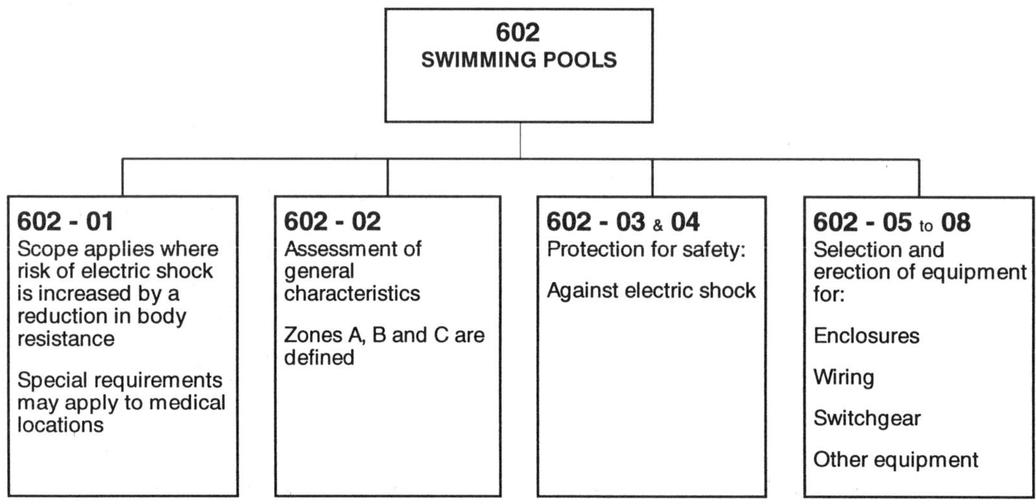

Special Point to Note

(a) Regulation 602 - 03 - 02 no longer requires a metal grid in a solid floor of zones B and C. However, where one is provided it shall be supplementary bonded.

Section 603

The conditions in a sauna steam room pose similar problems to the two previous situations with the added hazard of periods of high humidity which can cause deterioration of the installation with consequent increase in risks.

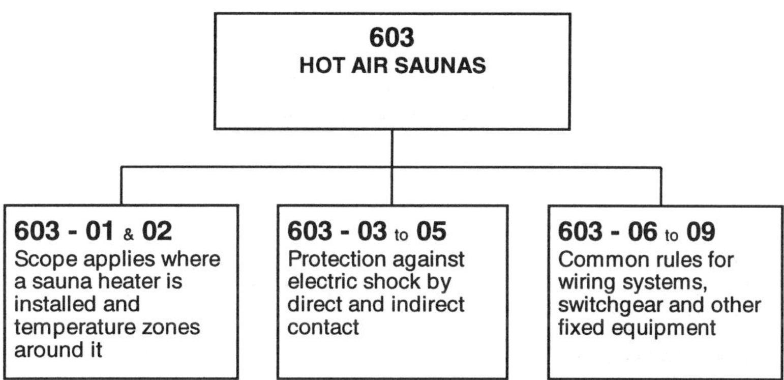

Section 604

Because the installations on a construction site are temporary they may be inherently less well protected and more exposed to damage during building operations. To give site operatives further protection from danger, this chapter sets out special requirements.

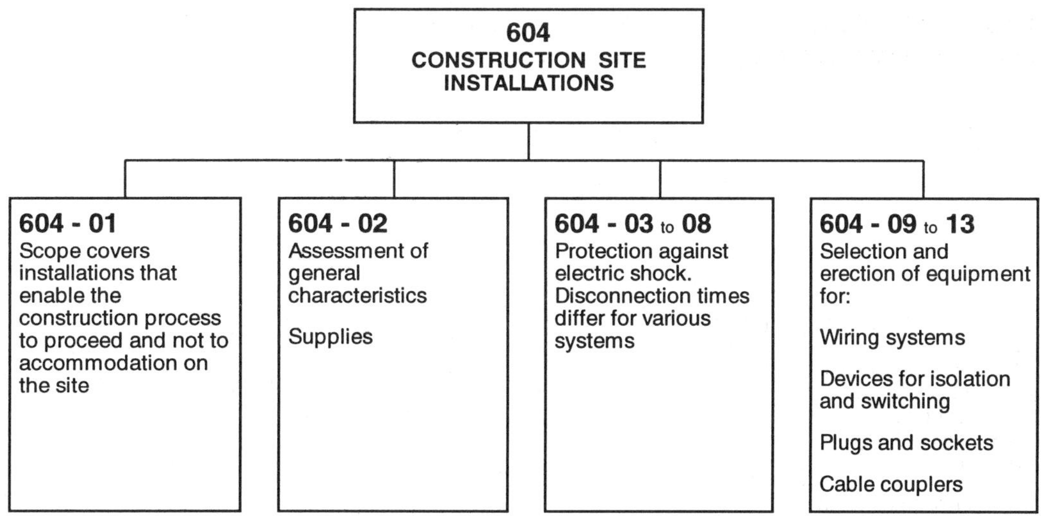

Section 605

Many installations in and around agricultural and horticultural buildings are in the open or virtually so. Also many of them are in areas where livestock is normally located. This section sets out the requirements to minimise the risk to persons and livestock.

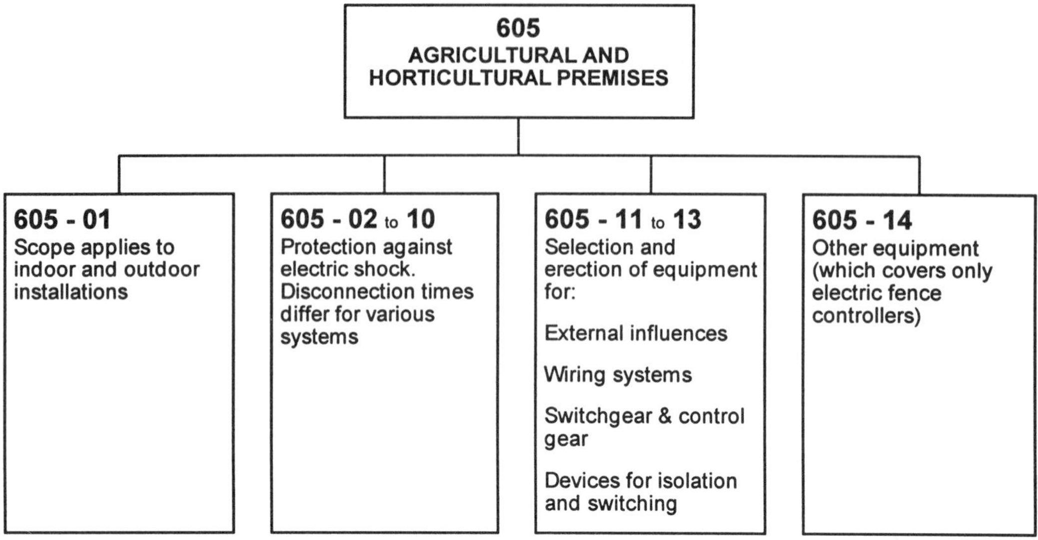

Section 606

Occasionally an operative is working in close proximity to conducting materials which restrict the person's movement, e.g. inside a metal vessel. Since the operative cannot move to avoid danger, this section sets down specific requirements to minimise the additional dangers arising from the location.

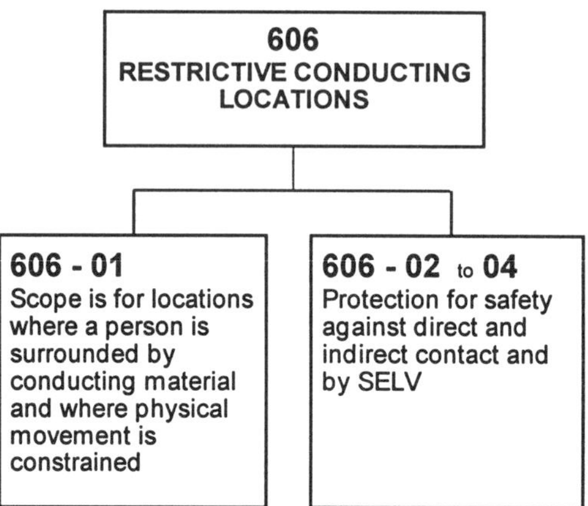

Section 607

Much of the electronic equipment now in use has an inherently high earth leakage current. To maintain safety under these conditions the earthing requirements of this chapter must be adhered to.

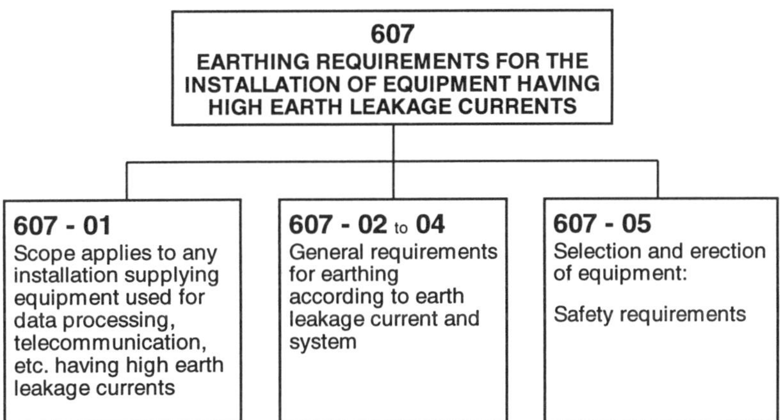

Section 608 Division One

Whilst this type of accommodation is basically like a house it is fed from a plug-in system and part of it may be subjected to a greater range of external influences. Thus additional requirements are set out in this chapter.

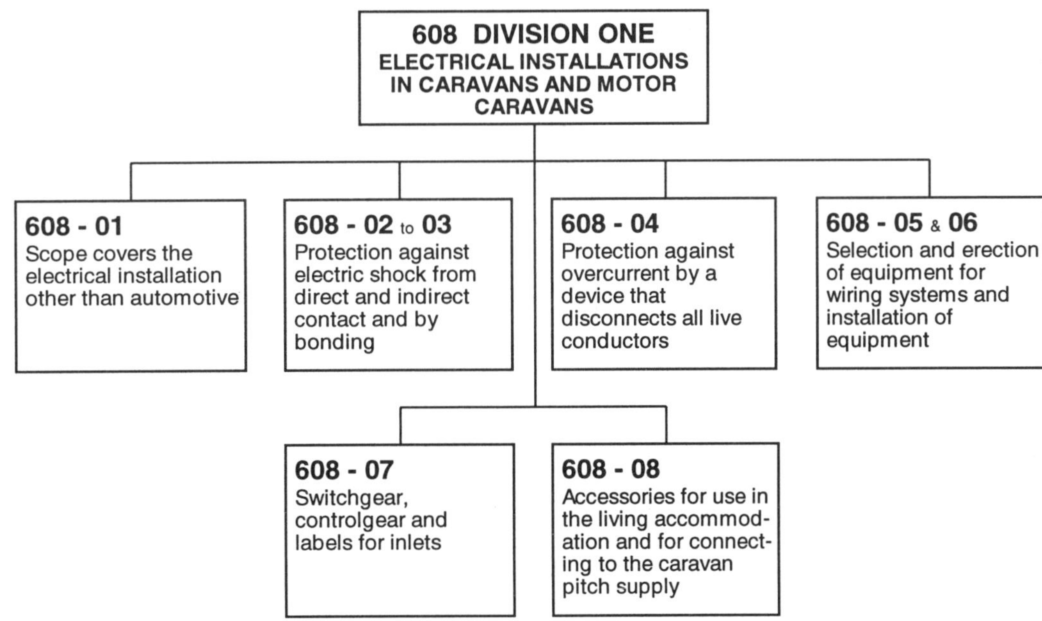

Section 608 Division Two

Since the installation in a caravan park may supply installations which are inferior or have been damaged during transit, this section sets out additional requirements to ensure the safety of the occupiers and restrict the influence of a fault on one circuit affecting other circuits or outlets.

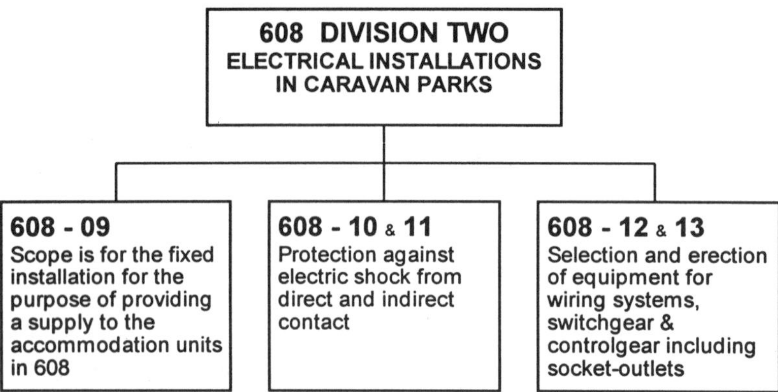

Section 611

This section details additional requirements to minimise the possibility of electric shock to persons using public places, other than buildings, where there are electrical installations.

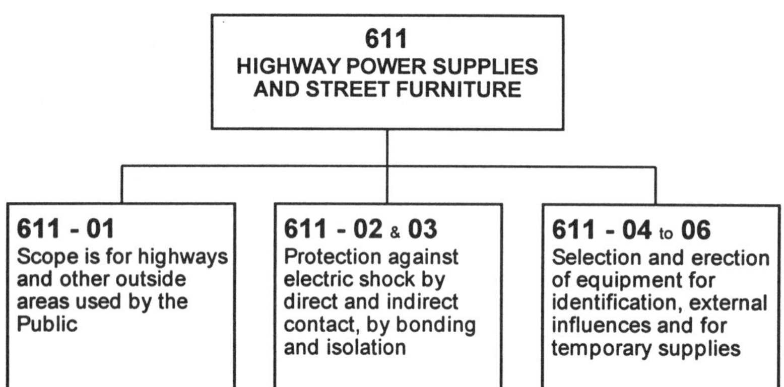

Inspection and Testing

After the design and the installation has been completed, there is a need to ensure that it meets the letter and the spirit of the Regulations. This is obtained by inspection, testing and recording as set out in this part.

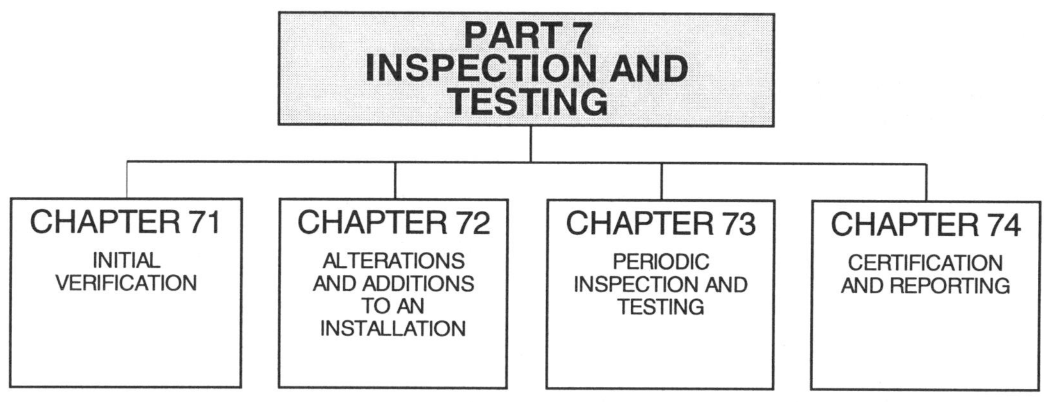

Chapter 71 — Initial Verification

Within this chapter are the requirements to ensure the new installation is in a suitable state to be put into service.

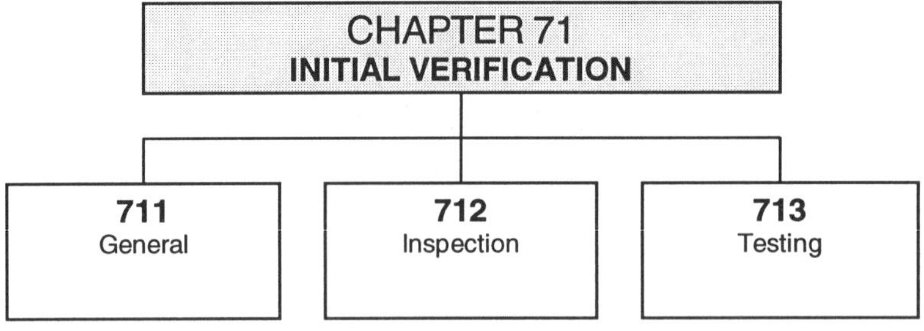

Sections 711 and 712

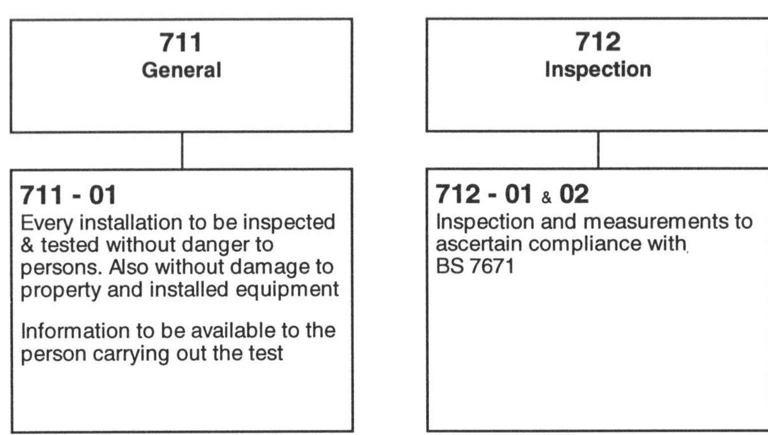

Section 713

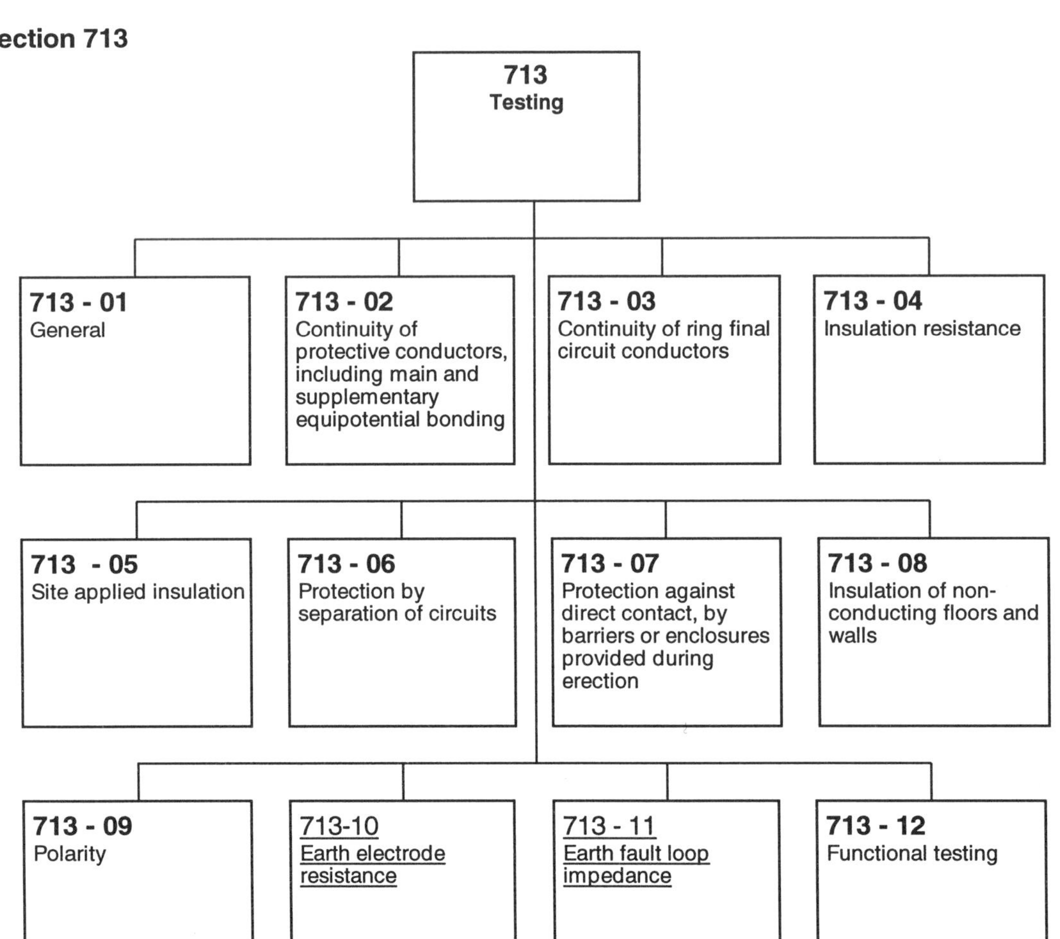

Chapter 72 — Alterations and Additions to Installations

A very short chapter containing two regulations, one with immense power in that it implies a need to check the safety of the existing installation after any further work (but not necessarily to the latest BS7671 requirements).

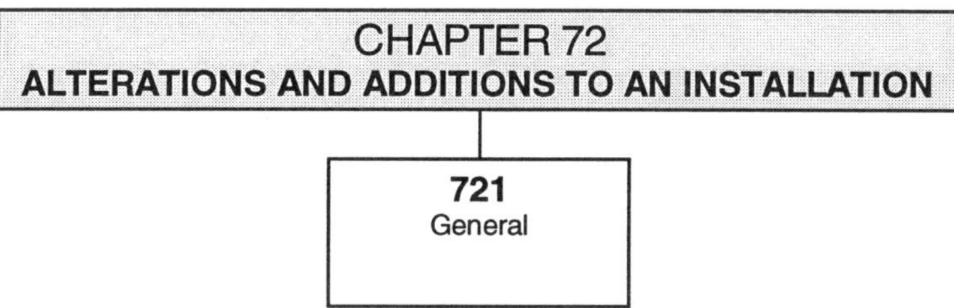

Chapter 73 — Periodic Inspection and Testing

Whilst the responsibility for initiating this is invested in the owner or user of the installation, the work will have to be carried out by a competent person. This chapter sets out the methods and requirements.

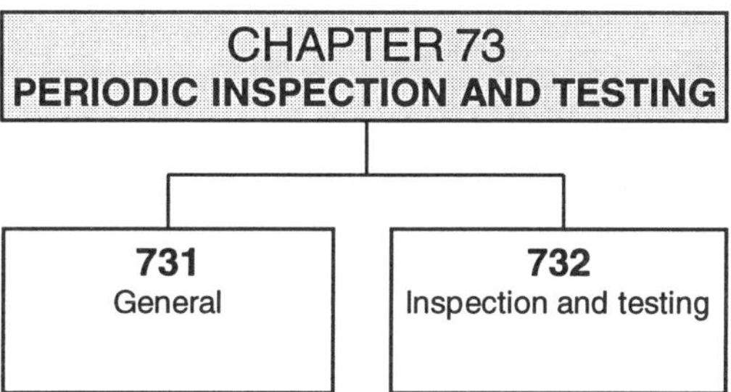

Chapter 74 — Certification and Reporting

A requirement for uniformity of information about installations and the applicable test results ensures that persons carrying out the tests adopt a consistent and professional approach.

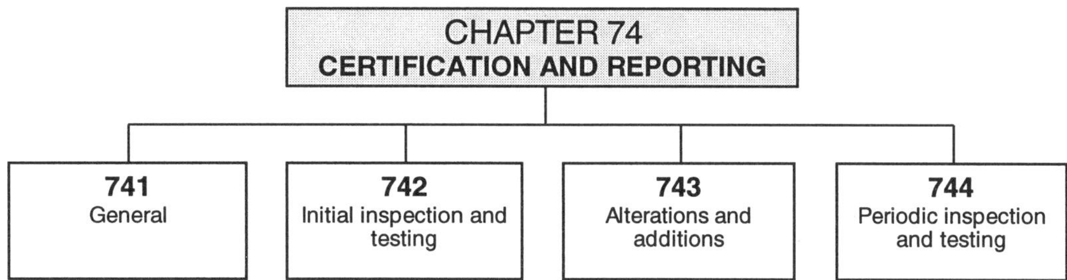

Special Points to Note

Regulation 741 - 01 - 01 requires a statement in respect of the design. Care must be taken before signing this if the design has not been carried out by the installer.

C

Topic Charts for Decision-making and Guidance

Topic Chart 1

AN INFORMATION CHART

Assessment of General Characteristics

Purposes, Supplies and Structure

The Maximum Demand (M.D.), of any installation, is the summation of all the installed loads less an estimation of the reduction in load due to diversity. Diversity is assessed from a knowledge of the operation of the system to determine what installed loads will not be on full load at the same time due to such factors as automatic time or temperature control, occupation patterns etc.

M.D. can be calculated on two bases:-

half hour for transformers, cables etc. and
short time for overload current, i.e. fuse or circuit breaker rating

For smaller installations guidance is given in IEE Guidance Notes, however, it must be recognized that whilst tables are all very well, there is no substitute for experience. Tables should be used with care and within the limits of one's own competence. For larger installations the assessment of M.D. and the diversity must be undertaken by a competent person.

Regulations in Sections 312 and 313 are intended to direct the designer (and in this context, the installer may also be acting as the designer) towards a consideration of the nature and characteristics of the supply on which the requirements for safety contained in Part 4 are selected. Examples of the information which will be required are the design levels for the prospective short circuit current, and the method of earthing which will affect the automatic disconnection times specified in Part 4.

Regulations in Section 314 contains the basic requirements for sub-division of the installation into manageable parts in order that any danger is avoided and inconvenience is minimized in the event of a fault, and to allow normal maintenance procedures to be carried out safely.

External Influences

No specific guidance is given on External Influences because the international work on this subject is still under consideration. It is important to ensure, however, that every item of equipment is suitable for the environment in which it is installed. Reference should also be made to Appendix 5 of BS 7671: 1992 which contains guidance on the factors to be considered when selecting equipment for particular locations.

Compatibility

The list of characteristics given later is not exhaustive and is only offered as a guide. The designer, installer, and the inspection and test engineer should ensure that all these aspects are taken into consideration.

The growing use of electronic equipment in general electrical installations (e.g. dimmer switches) merits particular attention, because in certain circumstances it can give rise to harmonics and peak voltages which can affect other equipment (see note on Page C/39).

TOPIC CHART 1 (Information) Assessment of General Characteristics

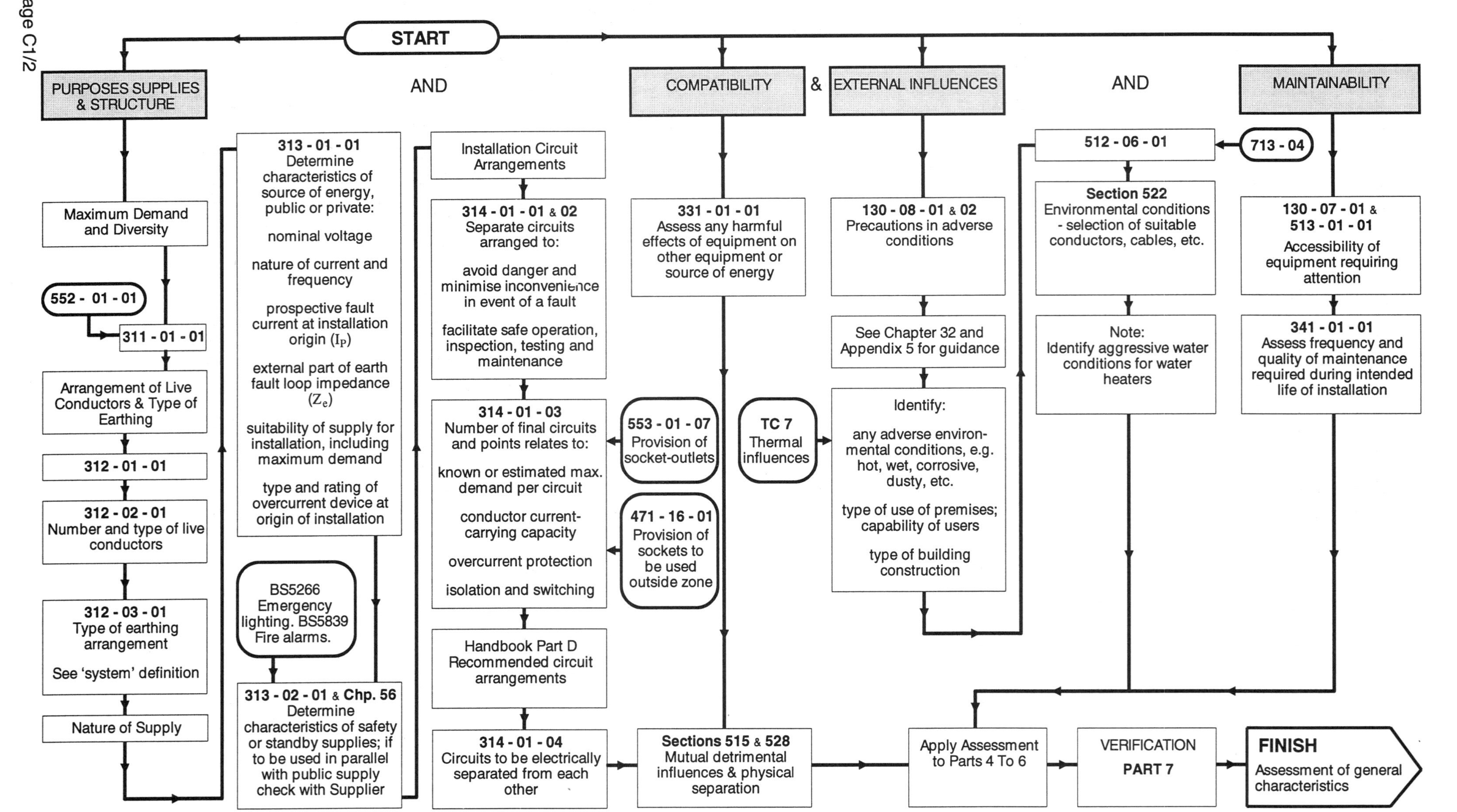

In addition, interference may be caused by transient disturbances generated by switching of any appliance containing capacitance or inductance.

ERA Report 97-0193 (Code of Practice for the Avoidance of Electrical Interference in Electronic Instrumentation and Systems) gives useful information in this area.

The following list contains some of the characteristics that affect other equipment or the supply.

1. transient overvoltages
2. rapidly fluctuating loads
3. starting currents
4. harmonic currents (such as with fluorescent lighting loads and variable speed drives)
5. d.c. feedback
6. high-frequency oscillations
7. earth leakage currents
8. the necessity for additional connections to an independent earth.

Increasing use of electronic equipment and controls makes the assessment of the harmful effect on the supply worthy of the fullest investigation and if in any doubt the situation should be discussed with the Supplier.

Maintainability

Chapter 34 draws attention to the necessity of considering the use or abuse to which the installation will be subjected during its life.

As well as the reliability and durability of the installation techniques, materials and equipment, the accessibility of all equipment must be considered so that regular inspection and maintenance can be readily and safely carried out.

Minimizing Danger and Inconvenience

Regulation 314 - 01 - 01 requires that circuits be arranged to avoid danger and minimize inconvenience in the event of a fault, and 314 - 01 - 02 requires consideration to be given to the consequences caused by the operation of a single protective device. The foreseeable dangers are more physical than electrical, and proper consideration of the difficulties which may arise from a faulty circuit will probably provide equally well against inconvenience as well as danger. It is fairly obvious that no matter how few lights are included in an installation, they should not all be on the same circuit, and a little thought will divide them to the best advantage e.g. in a small house with a staircase, the hall and landing should be on separate circuits, and it may be that a vertical rather than a horizontal division will provide some light in all parts of the premises with only one circuit operative.

Although recognised in BS 7671 by Regulation 531 - 04 - 01 for TT systems the provision of an RCD in place of an isolator to control the whole installation can lead to inconvenience and perhaps danger in the event of an earth fault. It is preferable to omit the lighting circuits from the protection of the RCD, bearing in mind that danger of shock is more likely to be from phase to neutral at a lampholder, and thus not protected by the RCD.

A split load consumer unit may be used so that a common isolator provides a single main control and separate parts of the unit can have RCD protection. For TN-S and TN-C-S systems RCD protection of socket outlet circuits only is the most common method used.

Consideration should however be given to 'essential' circuits such as those supplying deep-freeze units and in these circumstances RCD protection may not be appropriate.

Figure C1

Earthing Arrangements and Network Terminations Relating to the Various Systems

N.B. Systems IT and TN-C are not envisaged for general use and are, therefore, not shown.

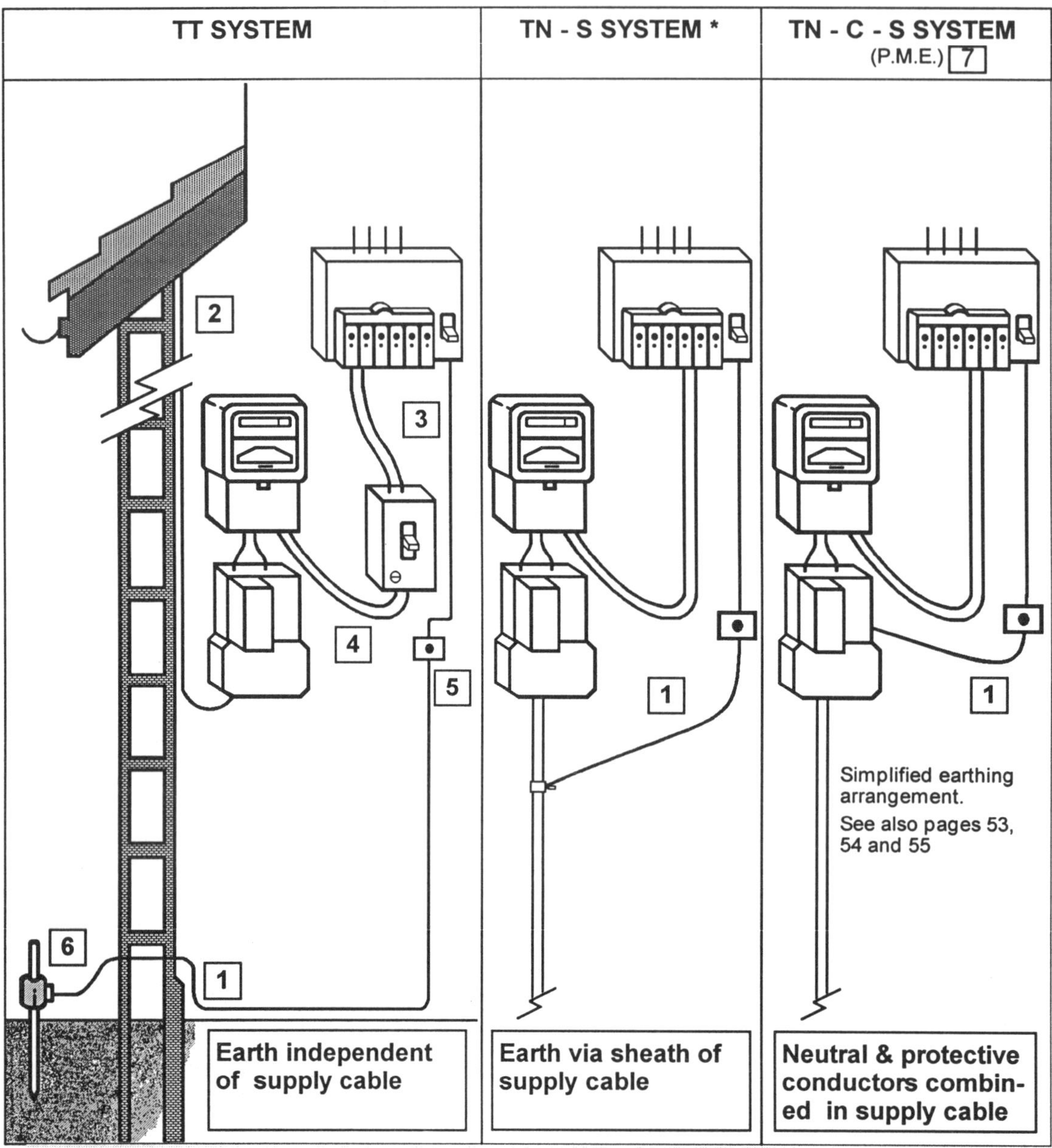

1 Earthing conductor
2 Cable from overhead supply
3 Earthing conductor connects directly to main earthing terminal
4 Cables not to be enclosed in metal conduit nor metal trunking as rcd does not protect against faults on supply side
5 Earthing conductor disconnecting terminal
6 Earth electrode - see Fig. C6(c) page C2/15
7 Where a PME service is, or may be, involved, there should be early discussion and agreement with the appropriate Supplier, most of which have produced explanatory booklets showing their requirements for such installations. (See definition of System in Part 2 of BS 7671: 1992) (rcd may be used for supplementary protection against direct contact).

Figure C2

Chapter 52
Environmental Conditions
Presence of water, dust, corrosive conditions

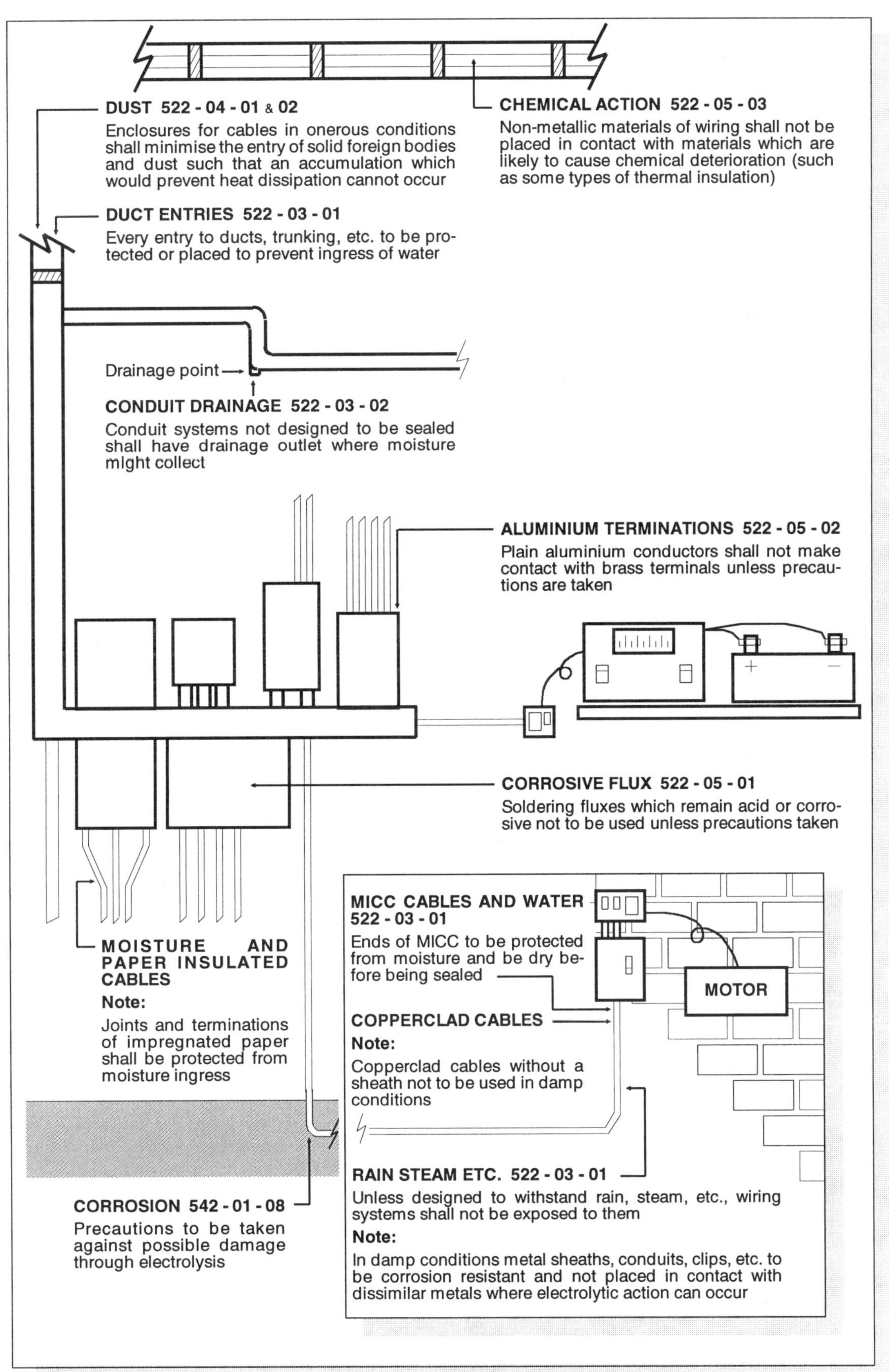

Figure C3 (a)

Chapter 52
Environmental Conditions
Ambient temperature and mechanical stresses
(See Figures C12(a) and (b) for mechanical supports)

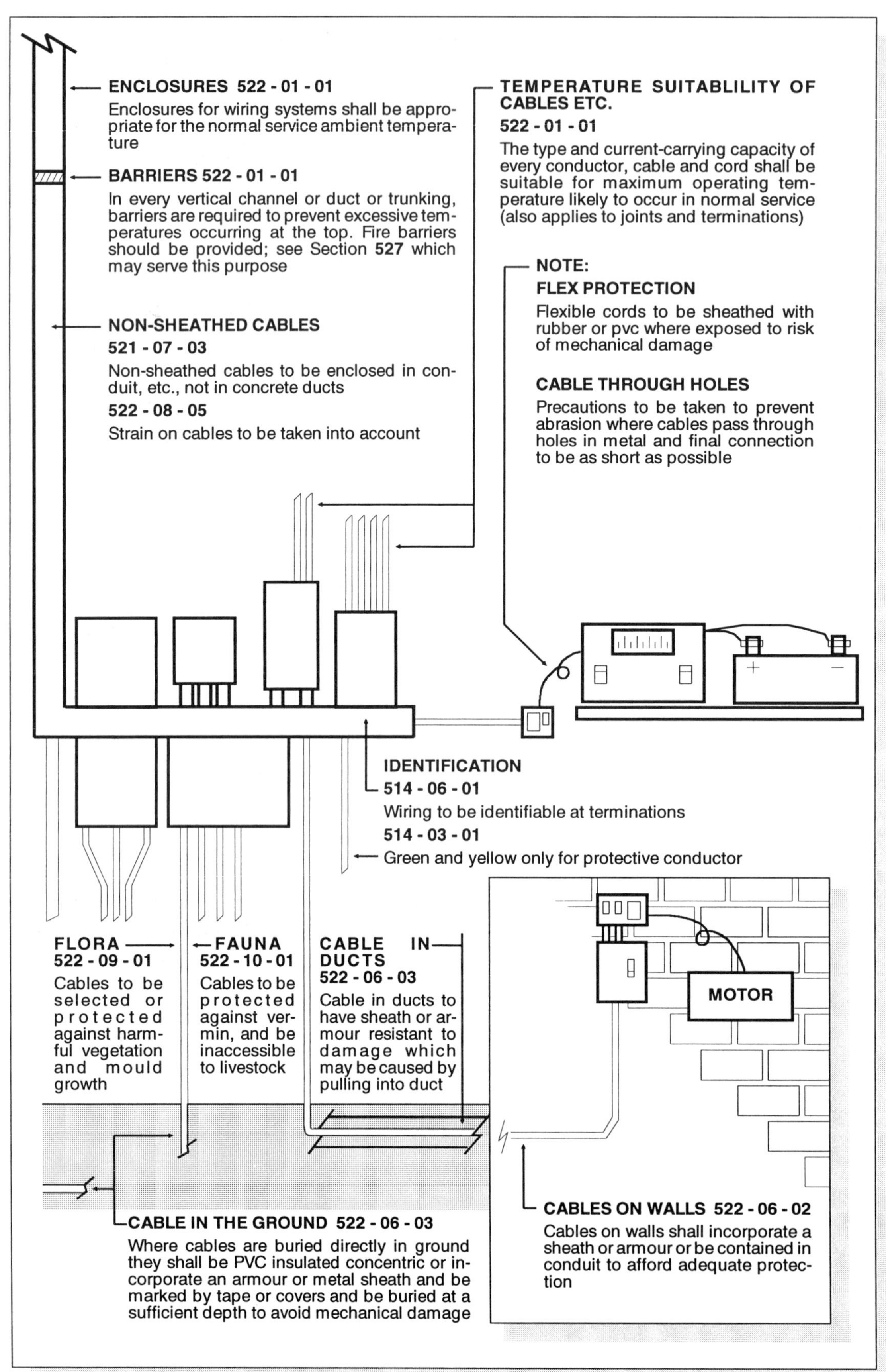

Figure C3(b)

Chapter 52
Environmental Conditions
Ambient temperature and mechanical stresses (continued)
(See Figures C12(a) and (b) for mechanical supports)

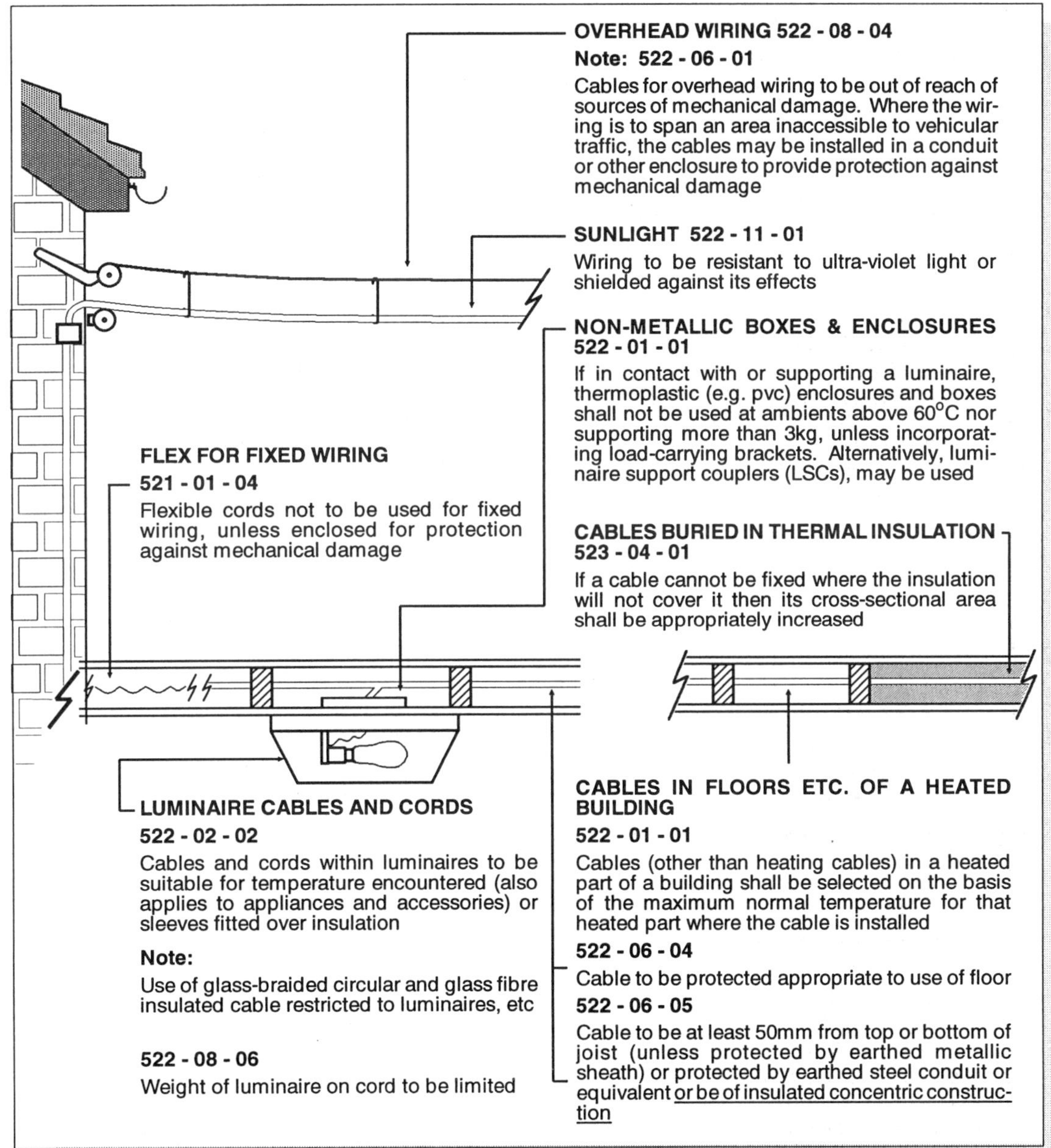

Figure C4

S.E.L.V. Circuits

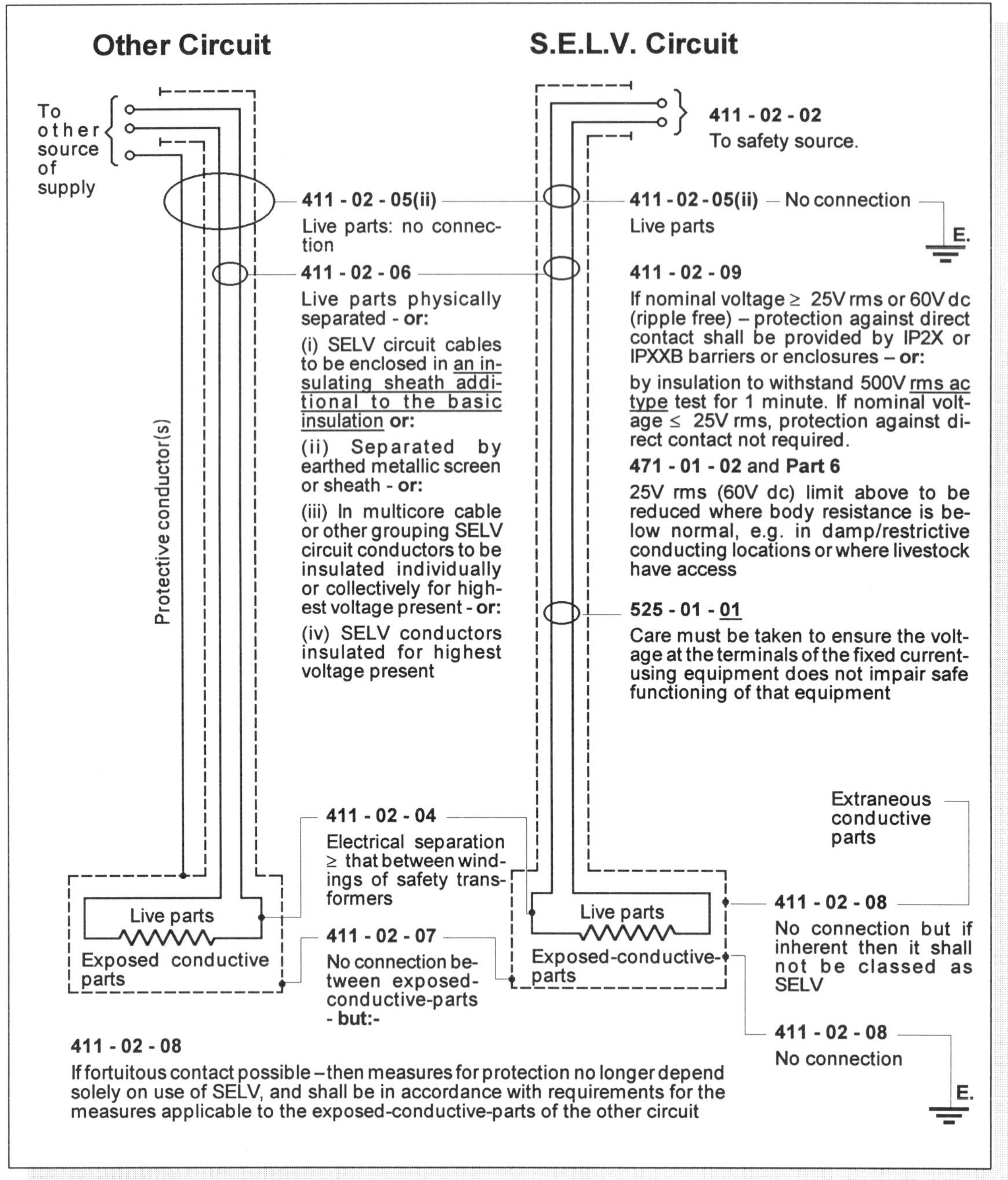

ITEM	411 - 02 - 10 Plugs and sockets for SELV circuits		For comparison, plugs and sockets for Extra-Low Voltage systems other than SELV. **471 - 14 - 01 to 06**		
SOCKET OUTLETS	Shall exclude plugs of other voltage systems in use in same premises	Shall not have protective conductor connection **except**	If no protective conductor connection	**471 - 14 - 06** Shall exclude plugs of other voltage systems in use in the same premises	**Note:** Protective conductor connection required
PLUGS	**411 - 02 - 10** Shall not enter a s/o of other voltage systems in use in same premises	Luminaire supporting couplers permitted subject to 471 - 14 - 06		**471 - 14 - 06** Shall not enter a s/o of other voltage systems in use in same premises	Luminaire supporting couplers permitted subject to 471 - 14 - 06

Note:

If all the requirements for SELV circuits cannot be met, e.g. an earth connection is required to the circuit, then the system becomes PELV and no additional protection is needed. If however, a safety source (see page 85) is not used then it is an Extra-Low Voltage system other than SELV, the exposed-conductive-parts of the circuit must be connected to the primary circuit protective conductor (471 - 14 - 03 to 05).

Figure C5

Segregation of Circuits – Band I and Band II

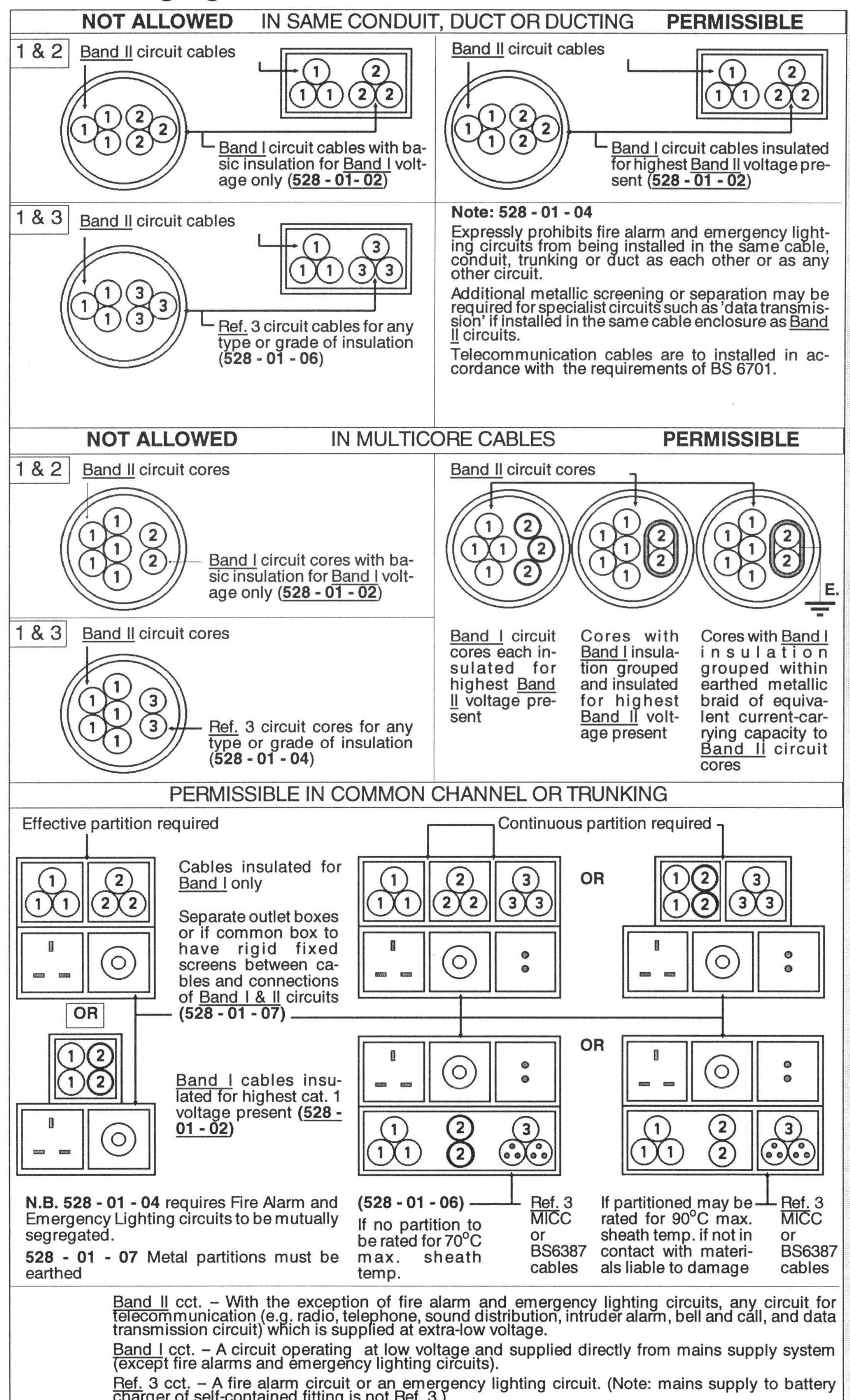

Topic Chart 2

A DECISION CHART

Protective Conductors and Earthing

An Exposed-Conductive-Part is a conductive part of equipment which can be touched, and which is not a live part, but which may become live under fault conditions. This definition refers to any metalwork which forms part of an electrical installation and applies to all metallic switchgear enclosures, machines, conduit, ducting, trunking, accessories, etc.

An Extraneous-Conductive-Part is a conductive part liable to introduce a potential, generally earth potential, and not forming part of the electrical installation. (See Note 1.)

A Protective Conductor is a conductor used for some measures of protection against electric shock, and intended for connecting together any of the following:

exposed-conductive-parts;

extraneous-conductive-parts;

the main earthing terminal;

earth electrode(s) or other means of earthing;

the earthed point of the source, or an artificial neutral.

This is a general definition which covers the following functions:

Bonding Conductor is a protective conductor for ensuring equipotential bonding.

Circuit Protective Conductor (cpc) is a protective conductor connecting exposed-conductive-parts of equipment to the main earthing terminal.

Earthing Conductor is a protective conductor connecting the main earthing terminal of an installation to an earth electrode or other means of earthing.

Note 1: The definition of Extraneous-Conductive-Part was altered during the validity of the 15th Edition to clarify its intention. The previous reference to "liable to introduce a potential", and the words "generally earth potential" replaced "including earth potential". If a conductive part is unlikely to introduce either earth or any other potential, it can be touched without danger and, therefore, does not need to be bonded, even though it may be simultaneously accessible with an exposed-conductive-part or an extraneous-conductive-part.

Continued...

See Part 2 of BS 7671: 1992 for the complete set of Definitions

TOPIC CHART 2 (Decision) Protective Conductors and Earthing

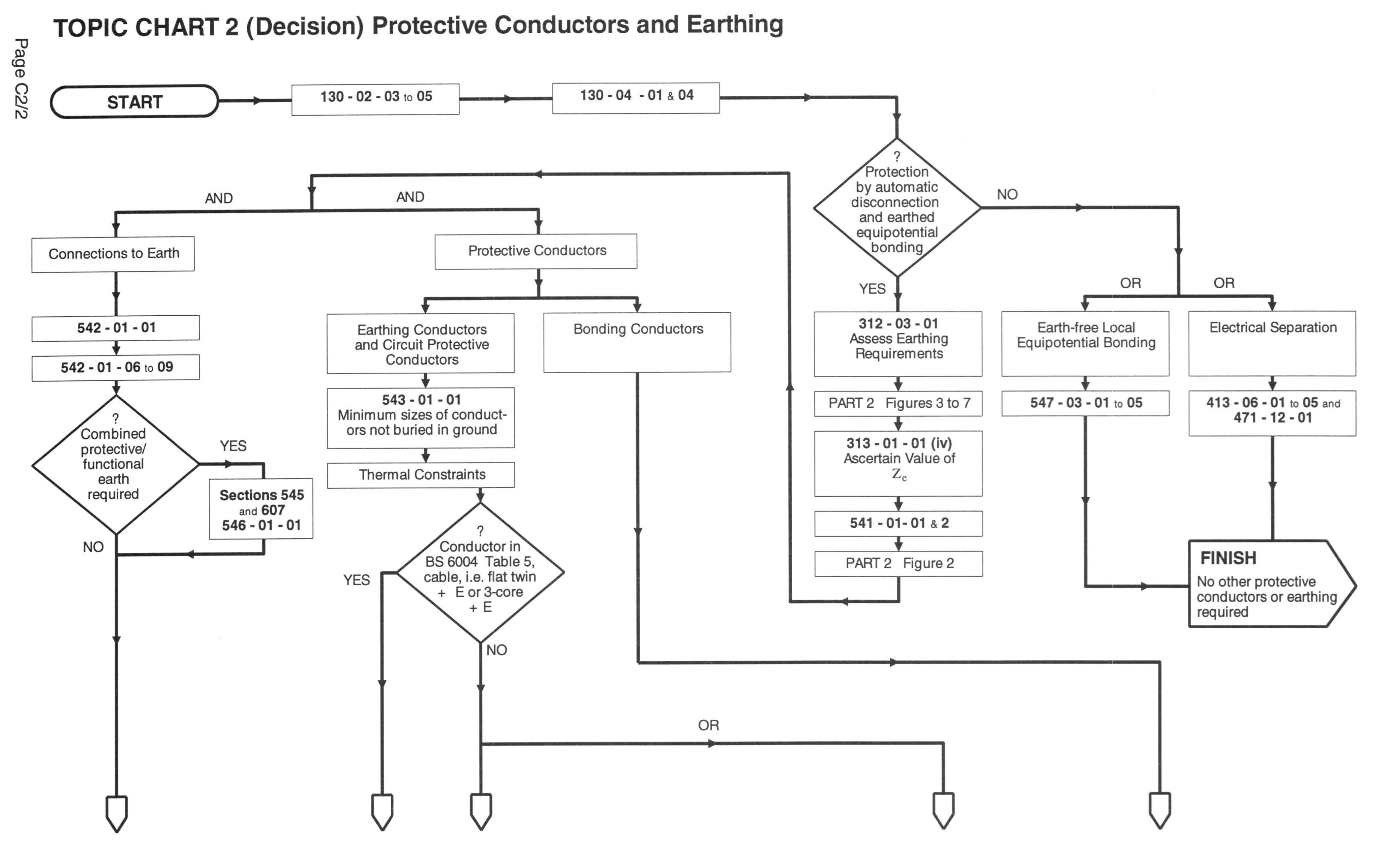

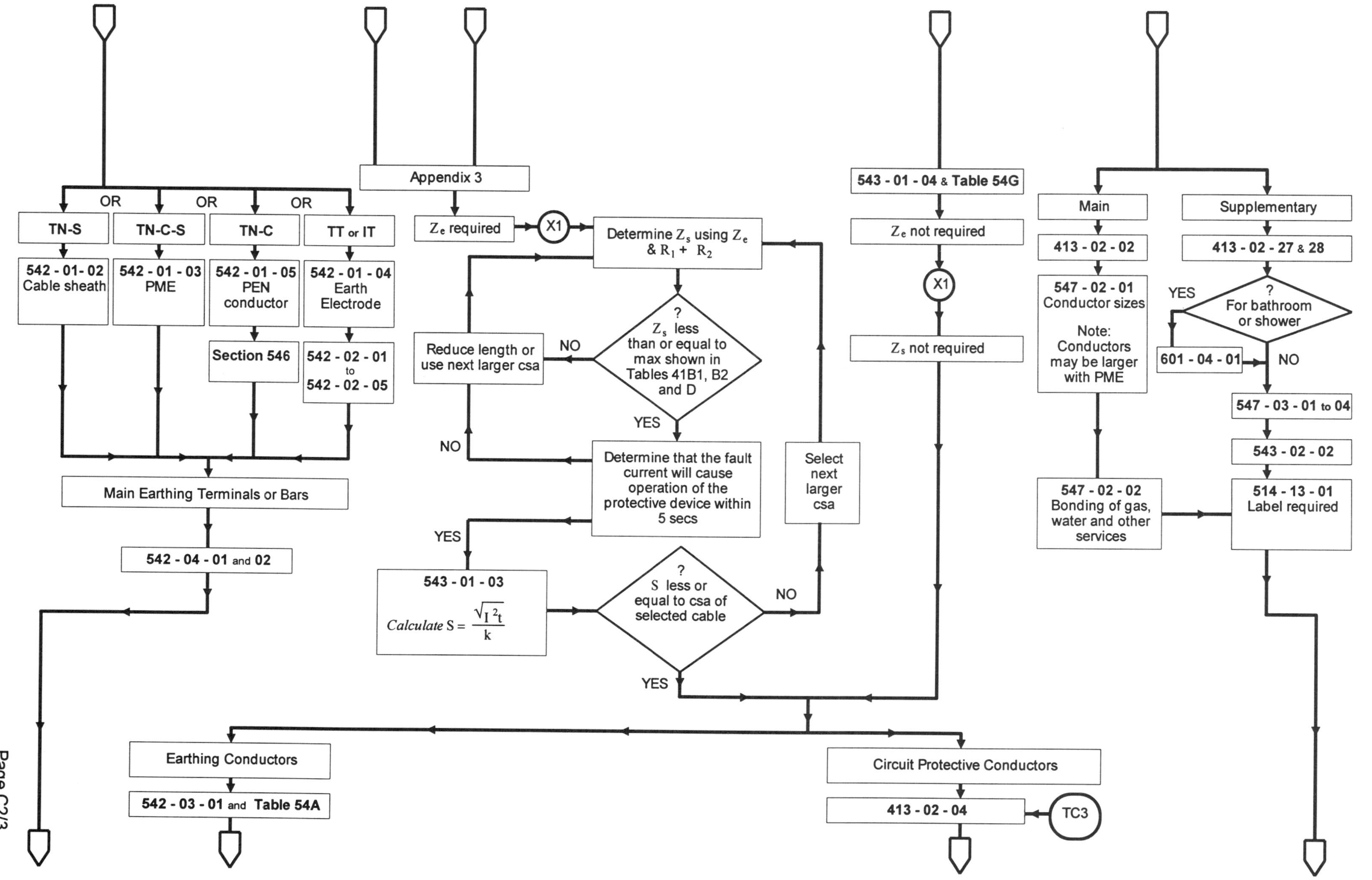

OR
TN-S
TN-C-S
TN-C
TT or IT
542 - 01 - 02
Cable sheath
542 - 01 - 03
PME
542 - 01 - 05
PEN
conductor
542 - 01 - 04
Earth
Electrode
Section 546
542 - 02 - 01
to
542 - 02 - 05
Main Earthing Terminals or Bars
542 - 04 - 01 and 02
Appendix 3
Z_e required
X1
Determine Z_s using Z_e & $R_1 + R_2$
?
Z_s less than or equal to max shown in Tables 41B1, B2 and D
NO
Reduce length or use next larger csa
YES
Determine that the fault current will cause operation of the protective device within 5 secs
543 - 01 - 03
Calculate $S = \frac{\sqrt{I^2t}}{k}$
?
S less or equal to csa of selected cable
Select next larger csa
543 - 01 - 04 & Table 54G
Z_e not required
Z_s not required
Main
413 - 02 - 02
547 - 02 - 01
Conductor sizes
Note:
Conductors may be larger with PME
547 - 02 - 02
Bonding of gas, water and other services
Supplementary
413 - 02 - 27 & 28
?
For bathroom or shower
601 - 04 - 01
547 - 03 - 01 to 04
543 - 02 - 02
514 - 13 - 01
Label required
Earthing Conductors
542 - 03 - 01 and Table 54A
Circuit Protective Conductors
413 - 02 - 04
TC3

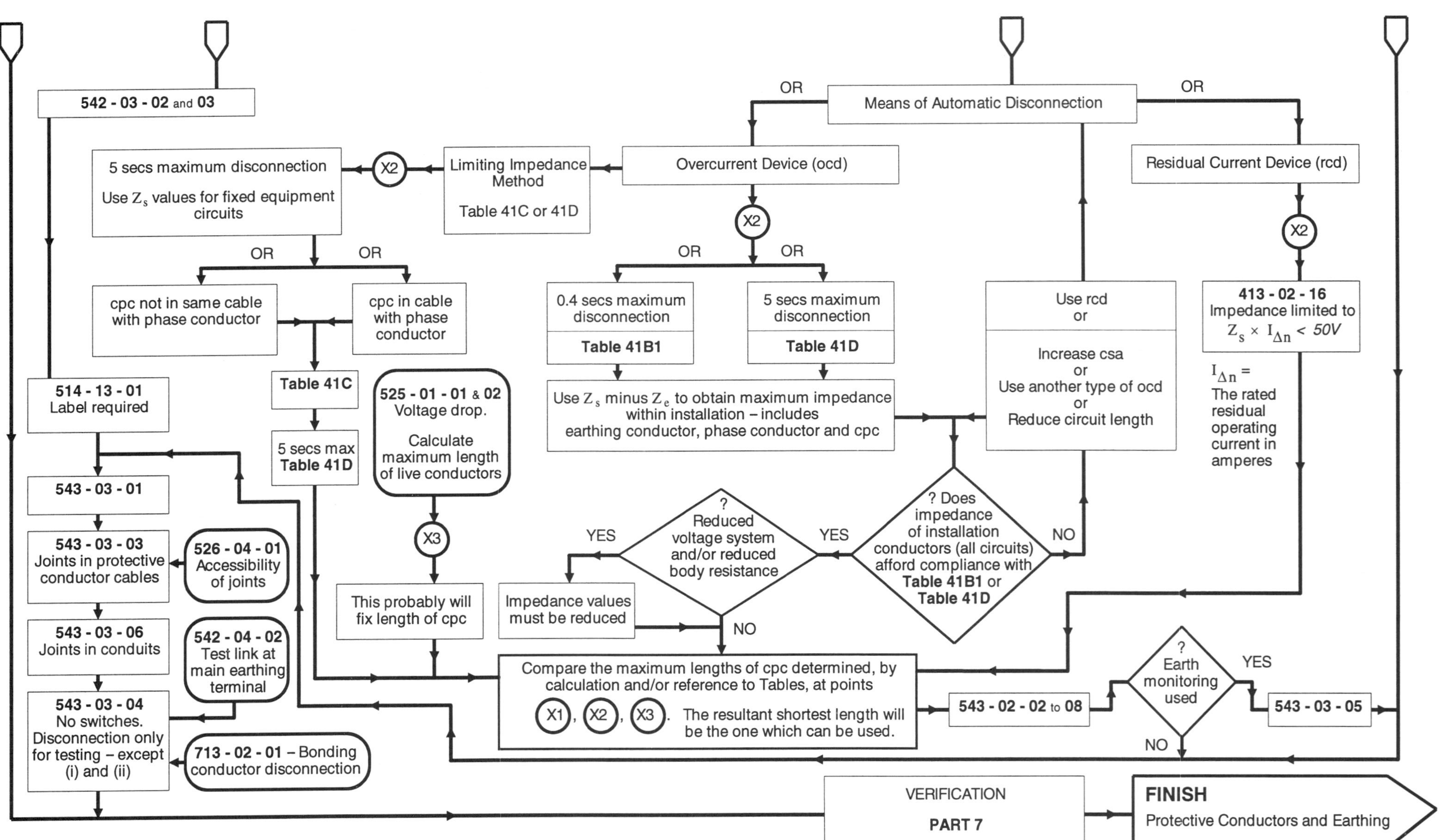
Means of Automatic Disconnection
OR
OR
Overcurrent Device (ocd)
Residual Current Device (rcd)
X2
X2
413 - 02 - 16
Impedance limited to $Z_s \times I_{\Delta n} < 50V$
$I_{\Delta n}$ =
The rated residual operating current in amperes
OR
OR
0.4 secs maximum disconnection
Table 41B1
5 secs maximum disconnection
Table 41D
Use Z_s minus Z_e to obtain maximum impedance within installation – includes earthing conductor, phase conductor and cpc
Use rcd
or
Increase csa
or
Use another type of ocd
or
Reduce circuit length
? Does impedance of installation conductors (all circuits) afford compliance with Table 41B1 or Table 41D
YES
NO
? Reduced voltage system and/or reduced body resistance
YES
NO
Impedance values must be reduced
Compare the maximum lengths of cpc determined, by calculation and/or reference to Tables, at points X1, X2, X3. The resultant shortest length will be the one which can be used.
543 - 02 - 02 to 08
? Earth monitoring used
YES
NO
543 - 03 - 05
VERIFICATION
PART 7
FINISH
Protective Conductors and Earthing
Limiting Impedance Method
Table 41C or 41D
X2
5 secs maximum disconnection
Use Z_s values for fixed equipment circuits
542 - 03 - 02 and 03
OR
OR
cpc not in same cable with phase conductor
cpc in cable with phase conductor
Table 41C
5 secs max
Table 41D
525 - 01 - 01 & 02
Voltage drop.
Calculate maximum length of live conductors
X3
This probably will fix length of cpc
514 - 13 - 01
Label required
543 - 03 - 01
543 - 03 - 03
Joints in protective conductor cables
526 - 04 - 01
Accessibility of joints
543 - 03 - 06
Joints in conduits
542 - 04 - 02
Test link at main earthing terminal
543 - 03 - 04
No switches. Disconnection only for testing – except (i) and (ii)
713 - 02 - 01 – Bonding conductor disconnection

Note 1 continued: For example, isolated door handles, coat hooks or shelf brackets are unlikely to introduce earth potential and are, therefore, not extraneous metal parts as defined, and do not need to be bonded.

On the other hand, metallic pipework for water, gas or oil is likely to be in contact with Earth, and must be deemed liable to introduce earth potential, so that appropriate main and perhaps supplementary bonding is required.

A generally acceptable test to establish whether an item is an extraneous-conductive-part is as follows:

Using a 500V d.c. insulation tester, measure the insulation resistance between the item and the main earthing terminal. If the resistance value is 0.25 megohm or greater, and inspection confirms that the resistance is unlikely to deteriorate, then the item can reasonably be considered not to be an extraneous-conductive-part.

The value of 0.25 megohm is based on the fact that 1mA flowing through the human body is not considered to be harmful, and this current will not be exceeded at the supply voltage with this resistance in circuit. Conversely if the insulation resistance test shows a negligible resistance, and this is confirmed by an ohm-meter as being 0.05 ohm or less, then the item is bonded and no further attention should be required.

BS 7671: 1992 regards supplementary equipotential bonding as being only required for ensuring the conditions for automatic disconnection (Regs. 413 - 02 - 08 to 14) are met, and for meeting the requirements of <u>Part 6</u>.

Where supplementary equipotential bonding is required, it must connect together the exposed-conductive-parts of equipment in the circuit concerned. These connections include the earthing contacts of socket outlets and any extraneous-conductive-parts, see Regulations 547 - 03 - 01 to 05 which detail the requirements of the bonding conductor and its installation.

In locations containing a bath tub or shower basin, there are different requirements, and it is not Regulation 413 - 02 - 15 which must be satisfied, but Regulation 601 - 04 - 01 which demands supplementary bonding for all simultaneously accessible <u>extraneous-conductive-parts and exposed-conductive-parts</u>. This applies even if the room has no facility for the connection of fixed current-using equipment, since without bonding dangerous voltages could exist between extraneous-conductive-parts in the event of an earth fault elsewhere in the installation.

Regulations 547 - 03 - 01 to 03 give the sizing of supplementary bonding conductors. The minimum is $4mm^2$ if unprotected for connecting to exposed-conductive-parts. For joining extraneous-conductive-parts the minimum size when protected is $2.5mm^2$.

"If in doubt, bond it!" is sometimes thought to be the universal answer, but it may not be. This approach can sometimes actually introduce hazards. The best example is that of metal windows and patio doors. Under fault conditions a voltage can appear on the main earthing terminal, and hence on every metal part connected to it. This is not a hazard within the equipotential zone, but if the metal windows/doors are bonded they will also be subjected to this raised voltage and would therefore present a shock hazard to a person entering from outside, or to a window cleaner, in contact with earth.

Note 2: A "main protective conductor" associated with, for example, a rising main feeding a sub-installation, may perform more than one specific function, and must therefore satisfy the requirements for each.

There are four main factors which may affect the determination of the size of a protective conductor:-

(a) Physical strength and protection against mechanical damage and corrosion;

(b) The capacity to carry earth fault current without thermal damage to itself and its surroundings;

(c) Impedance limitations imposed by the 'protection against indirect contact' requirements of Regulations 413 - 02 - 08 and 09;

(d) Voltage drop limits on associated live conductors where the protective conductor is a core within the same cable, or is within the same conduit, etc.

Factors (b), (c) and (d) may all result in limitations on the length of a protective conductor and it is necessary, therefore, to determine for each factor the related limitation on length.

The most onerous requirement of the three, i.e. the shortest of the three possible lengths, must be taken as the maximum permitted length of a protective conductor.

All Protective Conductors (Factor (a) above)

A main equipotential bonding conductor must have a cross-sectional area not less than 6mm^2 unless it is a TN-C-S system when the csa must not be less than 10mm^2.

If the supplementary bonding conductor is separate (i.e. if it is not part of cable, nor formed by conduit, ducting or trunking, nor within a wiring system enclosure) the cross-sectional area shall not in any event be less than:-

4 mm^2 if mechanical protection is not provided or

2.5mm^2 if mechanical protection is provided for connections between extraneous-conductive-parts.

Protective conductors of csa up to and including 10mm^2 must be of copper.

Circuit Protective Conductors

A. Thermal Constraints (Factor (b) above)

The cross-sectional area of protective conductors, other than equipotential bonding conductors, must be selected either by:-

(i) calculation in accordance with the equation in Regulation 543 - 01 - 03

$$S = \frac{\sqrt{I^2t}}{k} \text{ mm}^2$$

or (ii) application of Table 54G, in which case the cross-sectional area has a prescribed relationship to the associated phase conductor.

The application of (i) requires knowledge of the total earth fault loop impedance (Z_s) at the farthest point of a circuit. The nominal voltage divided by Z_s then gives the value of earth fault current at the farthest point. By using the time/current characteristics given in Appendix 3 of BS 7671: 1992, the time (t) relating to the value of earth fault current can be obtained.

These values of earth fault current and time are then used in the equation with the relevant value of k taken from Tables 54B to F, to obtain the minimum permitted cross-sectional area for the protective conductor.

This value is unlikely to be exactly a standard size and it is necessary, therefore, to select the nearest larger standard protective conductor.

Note: Reference should be made to BS 7430 which gives much more information on many aspects of earthing.

It must be emphasised that use of BS 6004 twin and cpc cables with reduced cpc requires application of the equation of Regulation 543 - 01 - 03. For values of I^2t related to the protective device see Appendix 3 of BS 7671: 1992.

The application of (ii) is very simple, as the protective conductor is related to the associated phase conductor in accordance with Table 54G, i.e. where all conductors are of the same material:-

phase conductor	protective conductor
$16mm^2$ or less	Same size
$25mm^2$ to $35mm^2$	$16mm^2$
over $35mm^2$	Half size (to nearest higher standard csa)

Where the protective conductor is not of the same material as the phase conductor, it must have the same conductance ratio to the phase conductor as that indicated above calculated by the formula in Table 54G.

This method automatically satisfies Factor (b) on page C2/6 and no knowledge of earth loop impedance or conductor length is required.

The method cannot be applied to BS 6004 twin and cpc cables with reduced cpc as, for conductors up to $16mm^2$, Table 54G requires the phase and protective conductor to be the same size.

If a circuit protective conductor is common to several circuits, its cross-sectional area is either to be calculated as described in sub-paragraph A(i) above, for the most unfavourable conditions of earth fault current and operating time in each of the associated circuits, or selected as described in sub-paragraph A(ii) according to the cross-sectional area of the largest associated phase conductor. Such common protective conductors may be metal conduit or trunking containing more than one circuit, or a separate conductor in conduit or trunking.

B. Indirect Contact Protection Constraints (Factor (c) on page C2/6)

Protective conductors must also comply with the 'protection against indirect contact' requirements of Section 413.

Whichever type of protection device is employed, it is necessary to comply with a limiting impedance for protective conductors which will often limit the length of a particular size of circuit protective conductor. Thus, it may be necessary to increase the protective conductor cross-sectional area to meet the requirements for shock protection.

The following alternative arrangements are available:

(i) For circuits requiring disconnection in 0.4 seconds or less, for compliance with Regulation 413 - 02 - 09, the tabulated values of maximum Z_s related to overcurrent protective devices in Table 41B1 and 41B2 apply. Although these values of impedance are identified for the disconnection of socket-outlet circuits in 0.4 seconds maximum, they must also be applied to circuits feeding fixed equipment in locations containing a bath tub or shower basin, which require disconnection in 0.4 seconds maximum (Regulation 601 - 04 - 02). In order to determine the earth fault loop impedance which can be tolerated within the installation (phase and protective conductors) it is necessary to deduct the external earth fault loop impedance (Z_e) from Z_s. Using the impedance per metre (for the actual cable being used, obtained from its manufacturer) for the phase and neutral conductors used in the installation the maximum length of cable can be calculated.

Fixed equipment supplied from a socket-outlet need not have a 0.4 second disconnection time if precautions are taken to prevent the use of the socket-outlet for supplying hand-held equipment.

This application requires the use of the specific maximum value of Z_e.

(ii) The application of Regulation 413 - 02 - 12 allows the disconnection time of a circuit supplying socket-outlets, portable equipment or hand-held Class I equipment to be increased to a maximum of 5 seconds. This entails compliance with the values of protective conductor impedance given in Table 41C according to the type and rating of the overcurrent protective device.

This offers advantages in the design of socket outlet circuits and should be used because:

(a) The degree of shock protection is unaffected if the fault is not of negligible impedance.

(b) The disconnection time can be increased from 0.4 seconds to 5 seconds.

(c) Larger values of external earth-loop impedance can be accepted.

Provided that the impedance of the protective conductor under fault conditions does not exceed the appropriate value in Table 41C, the shock voltage is limited to the internationally accepted "safe" value. Although the shock hazard is thus removed, the earth fault must not be allowed to persist indefinitely and it is also necessary, therefore, to comply with the relevant earth-loop impedance values in Table 41D to achieve disconnection within 5 seconds.

(iii) For fixed equipment circuits requiring disconnection in 5 seconds or less for compliance with Regulation 413 - 02 - 13, the tabulated values of maximum Z_S related to overcurrent protective devices in Table 41D apply. The maximum length of cable can be calculated as above.

This application requires the use of the specific maximum value of Z_e.

(iv) For installations or circuits protected by a residual current device, the limits of Z_S are determined by the application of Regulation 413 - 02 - 16.

The value of Z_S associated with these devices is usually of a higher order than the value associated with overcurrent protective devices. There should be no problem in achieving satisfactory impedance within an installation. The minimum sizes of protective conductors are likely to be determined by the relevant BS for the cable or by the overriding minimum requirements relating to Factor (a) set out on page C2/6.

Note: Where conditions of reduced body resistance exist, or in the presence of livestock, it will be necessary to reduce the impedance values. It is recommended that maximum impedance values should be reduced to not more than 25 per cent of those shown in Regulation 413 - 02 - 10 and 11, or other methods of protection should be employed.

C. Voltage Drop Constraints (Factor (d) on page C2/6)

Where a protective conductor has a fixed ratio of cross-sectional area to its associated phase conductor throughout its length, and where the two conductors run together throughout the installation (e.g. as in a BS 6004 twin and cpc cable, or in a pvc conduit installation, etc.) then the volt-drop limitation on the length of the phase conductor of a final circuit protected by an overcurrent device rated at 100A or less also limits the length of the associated protective conductor.

Where the 4.0% voltage drop applies, it is likely that the consequent limitation on length will be more onerous on protective conductors than will be the limitations on length imposed by the thermal constraints of Regulation 543 - 02 - 04 or the shock protection constraints of Regulations 413 - 02 - 08 or 09.

It is recommended, therefore, that voltage drop limitations should be determined before the other two possible limits on a protective conductor.

Circuit Protective Conductors

For domestic and similar installations, Regulation 413 - 02 - 12 is advantageous and, if it is possible to use multicore cables having a protective conductor with the same csa as the live conductors, then the adiabatic equation can be avoided by applying Regulation 543 - 01 - 04 and Table 54G.

Where protection against electric shock is provided by an overcurrent device the circuit protective conductor shall be incorporated in the same wiring system as the live conductors or in the immediate proximity of them (Regulation 544 - 01 - 01).

Earthing Conductors

In addition to meeting whichever of the foregoing requirements are relevant, earthing conductors are subject to additional requirements when buried in soil. Depending on whether or not an earthing conductor is protected against mechanical damage and/or protected against corrosion, it must satisfy Regulation 542 - 03 - 01 and the related Table 54A. Aluminium and copper clad aluminium must be protected against corrosion when used for earthing conductors with buried connections to earth electrodes. This is another requirement relating to Factor (a) on page C2/6.

Bonding Conductors

Requirements for sizing supplementary and main bonding conductors are contained in Section 547.

Whereas protective conductors are selected according to the level of earth fault current which they may carry, with equipotential bonding conductors it is virtually impossible to know exactly what earth fault current will flow.

Bonding conductors are selected, therefore, on the basis that the major proportion of an earth fault current will flow through the other protective conductors specifically intended to carry that current.

Main equipotential bonding conductors are required to have a cross-sectional area not less than half that of the related earthing conductor, subject to a minimum of 6mm^2 copper and a maximum of 25mm^2 copper, or equivalent conductivity.

Extraneous-conductive-parts are permitted to be used as protective conductors subject to Regulation 543 - 02 - 06 and all the requirements of Sub-section 542 - 02.

Important note:
Where a Protective Multiple Earthing (PME) service is provided by the Supplier, the main bonding connections must satisfy the requirements stated in the Electricity Supply Regulations 1988, as amended.

The cross-sectional area of the main bonding connections is to be related to the cross-sectional area of the supply neutral conductor (Regulation 547 - 02, Table 54H) but local public electricity supply network conditions may require a larger csa.

Copper equivalent csa of supply neutral	Minimum copper equivalent csa of main bonding conductor
35mm^2 or less	10mm^2
over 35mm^2 but not more than 50mm^2	16mm^2
over 50mm^2 but not more than 95mm^2	25mm^2
over 95mm^2 but not more than 150mm^2	35mm^2
over 150mm^2	50mm^2

The cross-sectional area of a supplementary bonding conductor is determined by a reference to its function and the associated circuit protective conductor as shown below. In each case, the minimum permitted cross-sectional area is 2.5mm^2 where protected against mechanical damage, or 4mm^2 where not so protected.

Supplementary bonding conductor connecting	Cross-sectional area not less than
Two exposed-conductive-parts (in bathrooms only)	Csa of smaller conductor connected to exposed-conductive-parts
Exposed and extraneous-conductive-parts	Half csa of protective conductor connected to exposed-conductive-part
Two extraneous-conductive-parts	2.5mm^2 protected; 4mm^2 not protected against mechanical damage
Two extraneous-conductive-parts, one of which is bonded to an exposed-conductive-part	Half csa of protective conductor connected to exposed-conductive-part

Regulation 547 - 03 - 05 permits the use of the circuit protective conductor within a flexible cord to provide the supplementary bonding connection.

Earthing Arrangements

Every installation must be provided with a main earthing terminal or bar for connecting;

(i) the circuit protective conductors,

(ii) the main bonding conductors, and

(iii) any other protective conductors

to the earthing conductor of the installation.

Provision must be made for a means of disconnecting the main earthing conductor from the means of earthing. This can often be conveniently combined with the main earthing bar. This provision must:

(i) be readily accessible,

(ii) be electrically reliable,

(iii) be mechanically sound, and

(iv) require the use of a tool for disconnection.

Attention is drawn to the need to disconnect the main equipotential bonding conductors for measurements of Z_e. See also page C4/4 Isolation and Switching.

The earthing conductor of the installation must be connected to the relevant means of earthing of the system, e.g. TN-S, TN-C-S, TT.

Note: Only the Supplier may make the connection to the PME earthing terminal.

Resistance of Conductors at Normal Operating Temperature

Regulation 413 - 02 - 05 requires that account should be taken of the increase of temperature and resistance of circuit conductors as a result of overcurrent and goes on to state that if the protective device complies with Appendix 3 of BS 7671 and the loop impedance of the circuit is complying with Regulations 413 - 02 - 10, 413 - 02 - 11 and 413 - 02 - 14, the circuit is deemed to comply with the requirements of this Regulation.

However, tables 41B1, 41B2, and 41D have a note which requires the values of Z_s should not be exceeded when the conductors are at their normal operating temperature.

This temperature is impossible to predict and it must be assumed that the normal operating temperature will not exceed the maximum allowable operating temperature of the particular type of cable.

Thus the value of Z_s given in tables 41B1, 41B2 and 41D must be assumed to apply when the cable is operating at its maximum permissible operating temperature.

When tests are carried out to ensure compliance with this requirement the cables will not be at their operating temperature but will be at the ambient temperature. A temperature co-efficient must therefore be applied to the measured Z_s before comparing it with the value given in the table.

The factor is calculated as follows:

$$1 + \left[\left(\begin{array}{l}\text{Maximum temperature}\\ \text{that a cable can}\\ \text{operate at}\end{array} - \begin{array}{l}\text{Ambient temperature used}\\ \text{for design or at which}\\ \text{tests are carried out}\end{array}\right) \times 0.004\right]$$

Thus for design based on or testing carried out at 20° C the factors for three of the most used cables are:

70° C PVC	1 + [(70° – 20°) × 0.004] = 1.2
85° C RUBBER	1 + [(85° – 20°) × 0.004] = 1.26
90° C THERMOSETTING	1 + [(90° – 20°) × 0.004] = 1.28

This used the simplified formula given in BS 6360 that the resistance–temperature coefficient is 0.004 for °C at 20° C for copper and aluminium conductors. Manufacturers will provide more accurate factors based on their own products.

The following example uses Manufacturers' typical values:

A circuit supplying socket outlets is wired with 2.5mm^2 2-core pvc-insulated and sheathed cable incorporating a 1.5mm^2 protective conductor. The circuit is protected by a 20A fuse to BS88 Pt 2. The external earth loop impenance, Z_e, is 0.8 ohm

Determine the maximum length of run affording compliance with Regulation 413 - 02 - 08 and 09 – as follows:

(a) Table 41B1 (a)

From table 41B1 (a), the maximum earth loop impedance, Z_s, associated with a 20A BS 88 fuse is 1.85 ohm.

As $Z_s = Z_e + (R_1 + R_2)$, where $(R_1 + R_2)$ is the resistance of the phase and protective conductors in series, the maximum permitted value of $(R_1 + R_2)$ is 1.85 – 0.8 = 1.05 ohm.

A typical value of $(R_1 + R_2)$/metre is 19.51 milliohm/metre.

The typical multiplier for pvc insulation is 1.2.

Therefore the value of $(R_1 + R_2)$/metre under fault conditions is 19.51 x 1.2 = 23.41 milliohm/metre.

The maximum length of run affording compliance with Table 41B1(a) is thus: $\dfrac{1.05}{0.02341} = 45$ metres

Resistance of Conductors under Fault Conditions

Where Regulation 434 - 03 - 03 applies i.e. conductors are protected against fault current only or it is required to calculate the size of protective conductors in accordance with Regulation 543 - 01 - 03 it will be necessary to check for thermal constraints under fault conditions by applying the adiabatic equation as follows:

$$t = \frac{k^2S^2}{I^2}$$

where t = duration of fault in seconds
I = fault current
S = csa of conductor
k = factor taking account of the heat capacity of the conductor

Example:

A 63 amp BS88 Pt 2 fuse provides overload protection for 16mm^2 copper PVC insulated conductors fed from busbars which are protected by 200 amp BS88 Pt 2 fuses. The 16mm^2 tails between the 200 amp and 63 amp fuses are 4.5 m long. The prospective fault current at the 63 amp fuse is 1kA. Establish whether the 200 amp fuses will provide fault current protection for the 16mm^2 tails.

$$t = \frac{k^2 S^2}{I^2}$$

$k =$ 115 from Table 43A BS7671
$S =$ 16mm^2
$I =$ 1000 amps

$$t = \frac{115^2 \times 16^2}{1000^2}$$

$t =$ 3.4 seconds

From Figure 3A (Appendix 3) BS7671 at 1000 amps a 200 amp BS88 fuse will disconnect in 7.5 seconds. Since the maximum permitted time (t) is only 3.4 secs it can be seen that the 200 amp fuse does *not* protect the tails against fault current.

A further calculation with a larger size of cable (25mm^2) would therefore be required.

$$t = \frac{115^2 \times 25^2}{1000^2}$$

$t =$ 8.2 seconds

Since the maximum duration time (t) has risen to 8.2 secs and the disconnection time is 7.5 secs 25mm^2 cable *is* protected against fault current.

Note: From Regulation 473 - 02 - 02 if the length of tails is limited to 3 metres and erected in such a manner as to minimise the risk of fire or danger to persons then the above calculation need not be applied.

Earthing Arrangements for Combined Protective and Functional Purposes

Functional earthing is defined as "earthing necessary for the proper functioning of equipment", and Regulation 331 - 01 - 01 requires that an assessment be made of possible harmful effects of this requirement. Regulation 542 - 01 - 06 allows that functional earthing may be either combined with, or separated from, protective earthing according to the requirements of the installation.

Section 607 deals with these matters in particular as they affect data processing equipment to BS 6204. There are two main divisions of earthing requirement. Regulation 545 - 01 - 01 requires that if there is a need for a "clean", low noise, earth, then this must be discussed and agreed with the equipment manufacturers, and it may be that combined protective and functional earthing is not possible.

However, if there is no such consideration then Regulations 607 - 02 - 01 to 07 (inclusive) detail the methods to be adopted, depending on whether the equipment has a normal leakage current of:

(a) less than 3.5mA, when no special requirements apply

(b) more than 3.5mA and less than 10mA when the connection of fixed wiring shall be either permanent or via BS EN 60309-2 plugs and sockets

(c) more than 10mA, when one or more of six methods of high integrity protective connections must be adopted, and the connection to the supply shall preferably be permanent although BS EN 60309-2 plugs and sockets are acceptable. Alternatively a monitoring system to BS 4444 (which disconnects the supply in accordance with regulation 413 - 02 - 01 in the event of a discontinuity in the protective conductor) may be used.

Regulations 607 - 02 - 03 and 607 - 03 - 01 give the requirements when rcd protection is provided, and it is pointed out that it may be necessary to supply such equipment from an isolating Transformer. This is a relatively specialised field, and a thorough study of the relevant Regulations is recommended.

Figure C6(a)

General Requirements for Protective Conductors

(for sizing see Figures C6(b) and D9)

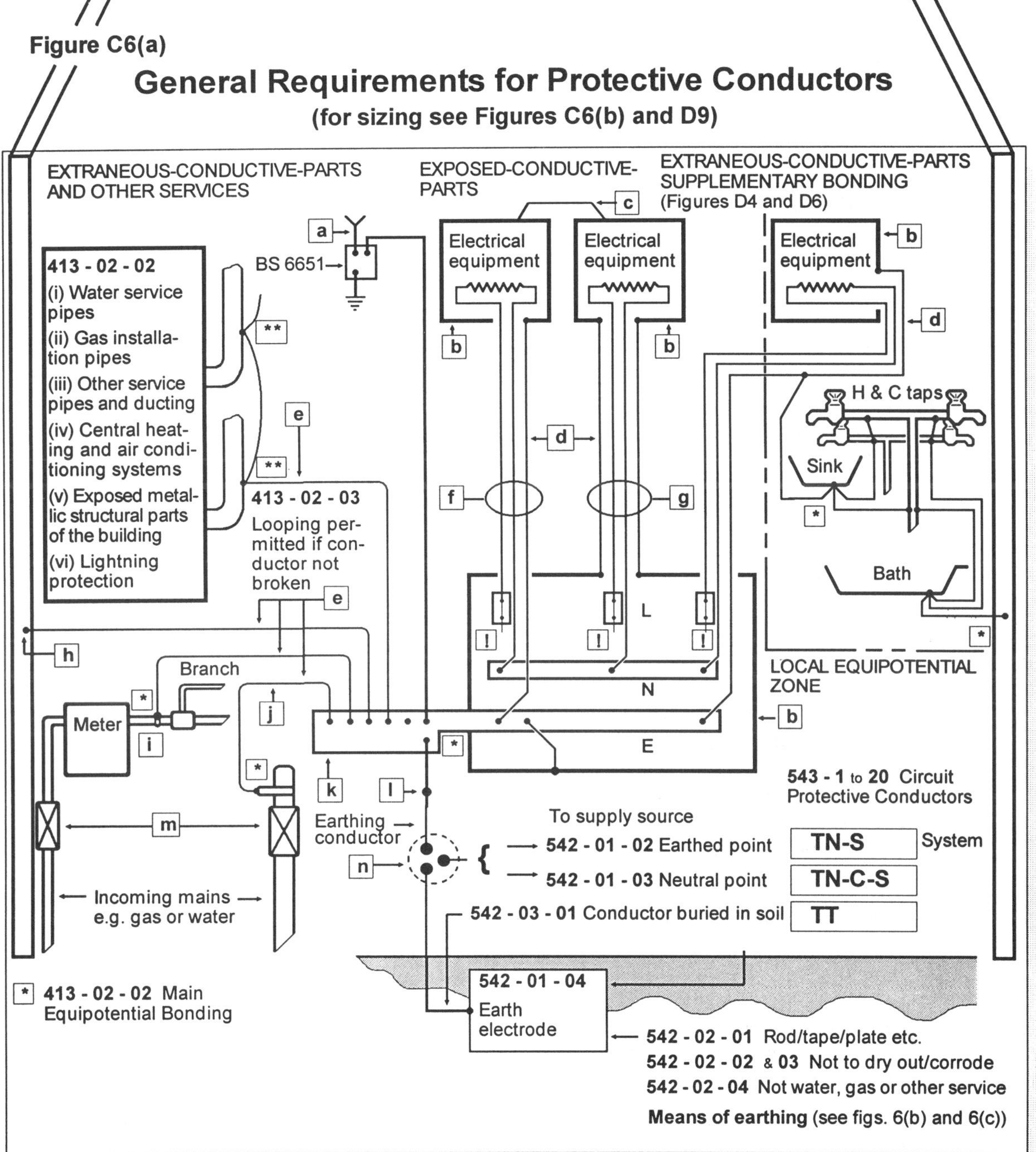

a. Lightning conductor
b. Exposed-conductive-parts
c. Supplementary bond
d. Protective conductor which can also be the supplementary bonding conductor if a flexible cord is used to feed the fixed equipment (547 - 03 - 05)
e. Main equipotential bonding (413 - 02 - 02)
f. Composite cable (543 - 02 - 02)
g. Conduit duct/trunking (543 - 02 - 02)
h. Exposed metallic parts of building structure
i. Maximum distance 600mm from service meter for equipotential bonding connection (547 - 02 - 02)
j. Not Al or CuAl if subject to corrosion
k. Main earth terminal (542 - 04 - 01)
l. Accessible means of disconnection by tool only (542 - 04 - 02)
m. Insulating section (if any) all bonding to be on consumer's side (547 - 02 - 02)
n. Connection of earthing conductor to means of earthing protected against corrosion (542 - 03 - 03)

* Accessible connection
543 - 03 - 03 Accessible connection and
514 - 13 - 01 Warning notice
SAFETY ELECTRICAL CONNECTION
DO NOT REMOVE

** Accessible and unbroken connection

! Protective devices. Characteristics and earth fault loop impedance (Z_s) of circuits to be coordinated so that:

– socket-outlet circuits are disconnected in 0.4 seconds or less
– fixed equipment circuits are disconnected in 5.0 seconds or less
(Regs. 413 - 02 - 04 to 09 and Tables 41B1 to 41D)

and for rcds Z_s x tripping current should be less than 50 volts (413 - 02 - 16)

Note: BS 951 compliance necessary for all connections clamped to pipes

Figure C6(b)

Sizing: Protective Conductors; Earthing Conductors; Main Equipotential Bonding Conductors; Circuit Protective Conductors

543 - 01 - 01 and **02**

Calculation or selection of protective conductor cross-sectional area – equipotential bonds excluded

543 - 01 - 03

To be determined as:-

$S(csa)mm^2$ of protective conductor $= \frac{\sqrt{I^2t}}{k}$

t = disconnection time

k = Material factor from Reg. 543 - 01 - 03 of the Regulations

I = Fault current

OR

NOTE:

BS 6651 (Protection of Structures against Lightning) requires that a main equipotential bond be extended to the lightning conductor, via as short a route as possible to the conductor side of the test clamp. BS 6651 also requires bonding to liftshaft metal, e.g. guides, handrails, ladders, etc. at top and bottom of the shaft and additionally at intervals of not more than 20m which is more than 'extraneous metal' would require for 16th Edition compliance.

See Tables 41B1, 41B2, 41C and 41D for calculated values of maximum earth loop impedance relating to BS 6004 cables which take into consideration the maximum value of t, i.e. 0.4 or 5 seconds.

Size to be related to that of phase conductor (S mm^2) – see table 54G

Where S is equal or less than $16mm^2$ protective conductor to be equal to S

S is greater than $16mm^2$ and less or equal to $35mm^2$ protective conductor to be $16mm^2$

S is greater than $35mm^2$ protective conductor to be half the phase conductor csa

If protective and phase conductor are not of same material then conductance of protective conductor to be not less than that conductance it would have if it were of the same material and selected as stated above.

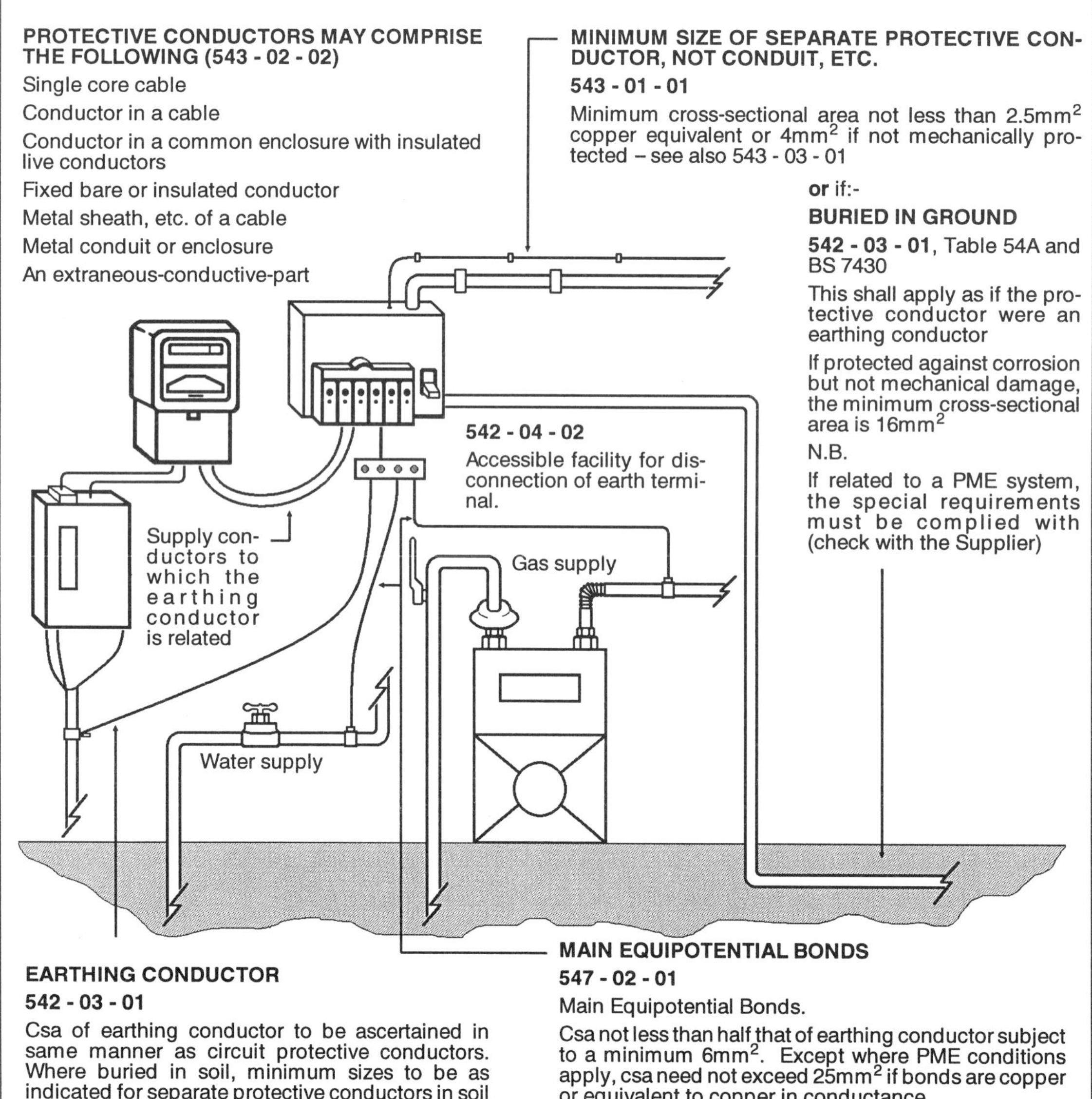

PROTECTIVE CONDUCTORS MAY COMPRISE THE FOLLOWING (543 - 02 - 02)

Single core cable

Conductor in a cable

Conductor in a common enclosure with insulated live conductors

Fixed bare or insulated conductor

Metal sheath, etc. of a cable

Metal conduit or enclosure

An extraneous-conductive-part

MINIMUM SIZE OF SEPARATE PROTECTIVE CONDUCTOR, NOT CONDUIT, ETC.

543 - 01 - 01

Minimum cross-sectional area not less than $2.5mm^2$ copper equivalent or $4mm^2$ if not mechanically protected – see also 543 - 03 - 01

or if:-

BURIED IN GROUND

542 - 03 - 01, Table 54A and BS 7430

This shall apply as if the protective conductor were an earthing conductor

If protected against corrosion but not mechanical damage, the minimum cross-sectional area is $16mm^2$

N.B.

If related to a PME system, the special requirements must be complied with (check with the Supplier)

EARTHING CONDUCTOR

542 - 03 - 01

Csa of earthing conductor to be ascertained in same manner as circuit protective conductors. Where buried in soil, minimum sizes to be as indicated for separate protective conductors in soil

MAIN EQUIPOTENTIAL BONDS

547 - 02 - 01

Main Equipotential Bonds.

Csa not less than half that of earthing conductor subject to a minimum $6mm^2$. Except where PME conditions apply, csa need not exceed $25mm^2$ if bonds are copper or equivalent to copper in conductance

Figure C6(c)

The Seven Recognised Types of Earth Electrode (542 - 02 - 01)

ALUMINIUM

Aluminium and copper-clad not to be used for final connection to earth electrodes where corrosion may occur

CONNECTION

542 - 03 - 03

Connection to earth electrode to be labelled, soundly made, protected against corrosion and electrically and mechanically satisfactory (See 514 - 09 - 01 for identification)

MATERIALS & CONSTRUCTION

542 - 02 - 03

Materials used and the construction must be such as to withstand damage

CORROSION

542 - 02 - 03

Design of earthing arrangements to take account of possible increase in earth resistance due to corrosion

DRYING / FREEZING

542 - 02 - 02

Type and embedded depth must be such that soil drying or freezing will not increase resistance unacceptably

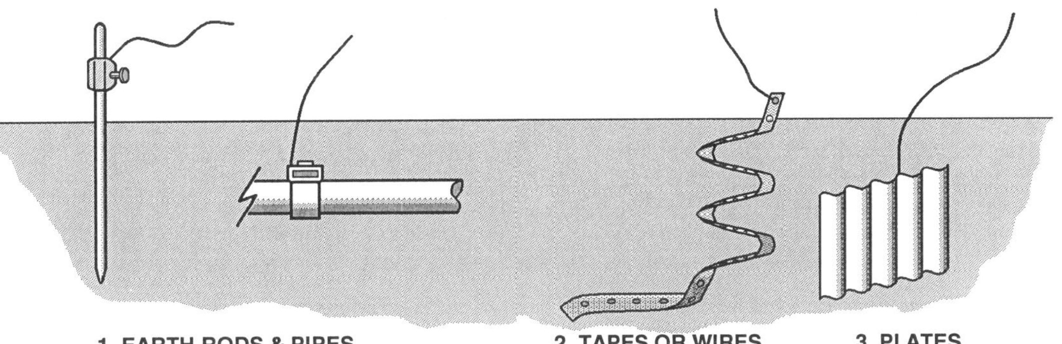

1. EARTH RODS & PIPES

542 - 02 - 04

Gas and water services not to be used as protective earth electrodes

2. TAPES OR WIRES

3. PLATES

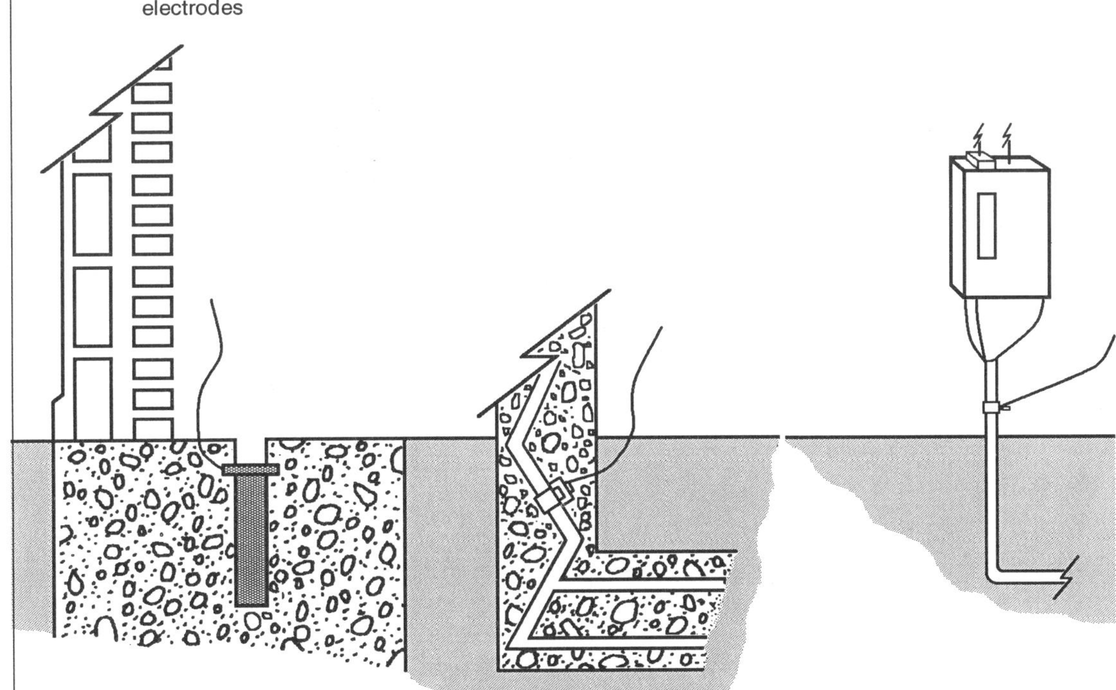

4. UNDERGROUND STRUCTURAL METALWORK EMBEDDED IN FOUNDATIONS

See BS 7430 for dimensional details of earth electrodes

5. WELDED METAL REINFORCEMENT OF CONCRETE (NOT PRE-STRESSED)

7. OTHER SUITABLE UNDERGROUND METALWORK (not illustrated)

6. LEAD OR METALLIC COVERINGS OF CABLES

542 - 03 - 01

lead or metallic covering must be in effective contact with the ground. Consent of owner to be obtained. Arrangements for installation owner to be notified of effective changes which may be proposed to be made to the cables.

Topic Chart 3

AN INFORMATION CHART

Protection against Electric Shock

Direct Contact. Contact of persons or livestock with live parts.

Indirect Contact. Contact of persons or livestock with exposed-conductive-parts which have become live under fault conditions.

Live Part. A conductor or conductive part intended to be energised in normal use, including a neutral conductor but, by convention, not a PEN conductor.

SELV. An Extra-low voltage system which is electrically separated from earth and from other systems in such a way that a single fault cannot give rise to the risk of electric shock.

PELV (protective extra-low voltage). An extra-low voltage system which is not electrically separated from earth, but which otherwise satisfies all the requirements for SELV.

Hazardous-live-part. A live part which can give, under certain conditions of external influence, an electric shock.

See Part 2 of BS 7671: 1992 for the complete set of Definitions

TOPIC CHART 3 (Information) Protection against Electric Shock

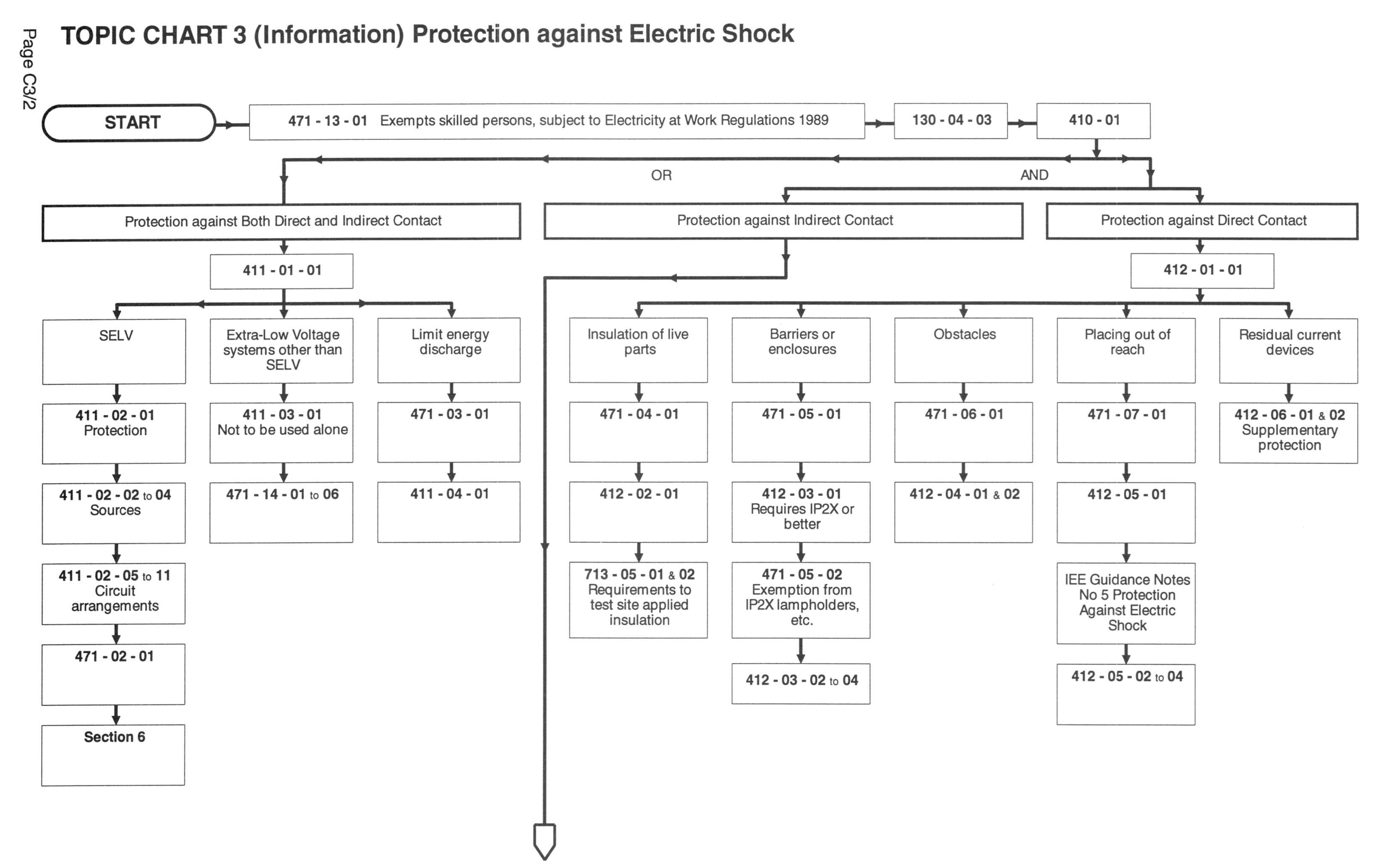

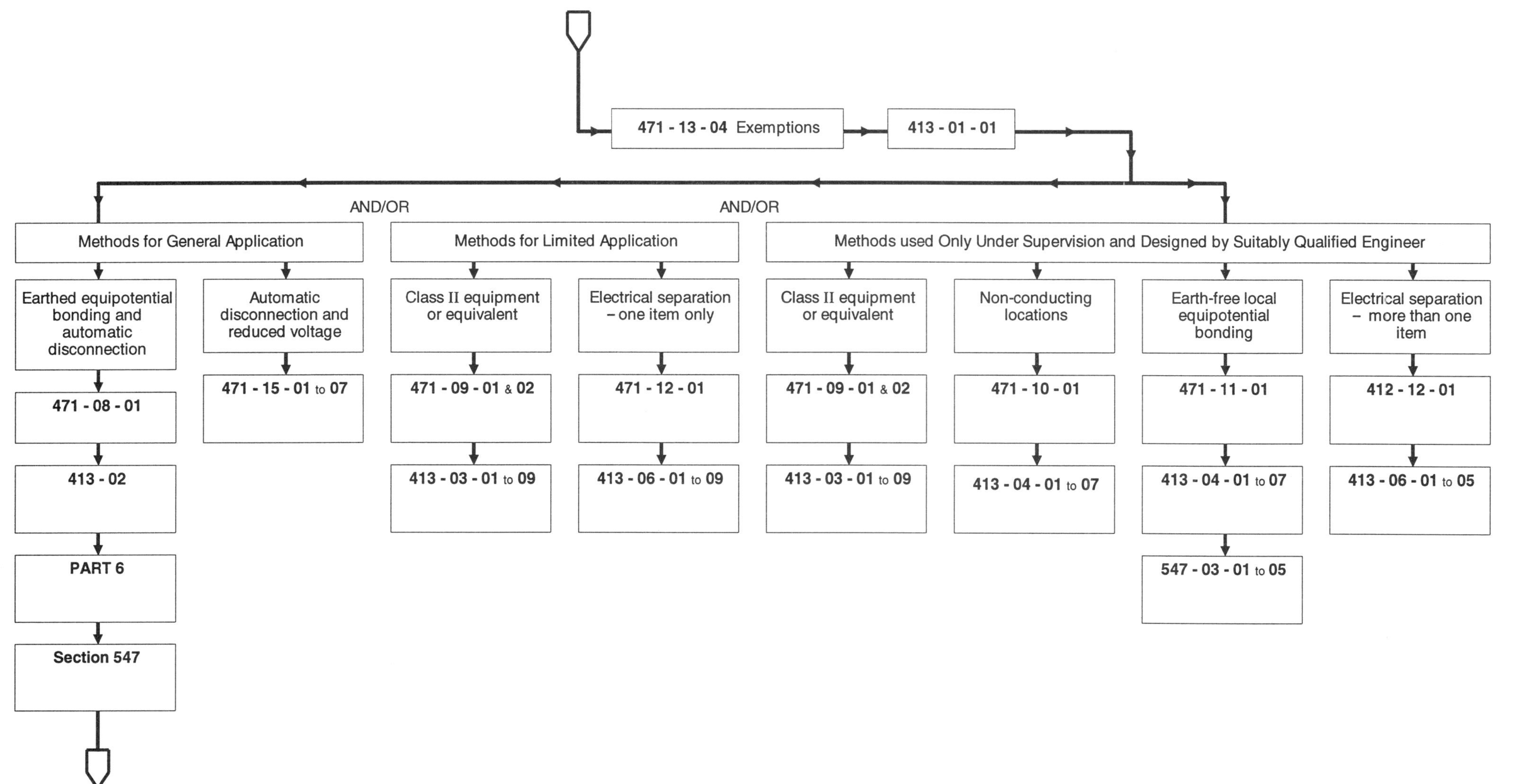

471 - 13 - 04 Exemptions
413 - 01 - 01
AND/OR
AND/OR
Methods for General Application
Methods for Limited Application
Methods used Only Under Supervision and Designed by Suitably Qualified Engineer
Earthed equipotential bonding and automatic disconnection
471 - 08 - 01
413 - 02
PART 6
Section 547
Automatic disconnection and reduced voltage
471 - 15 - 01 to 07
Class II equipment or equivalent
471 - 09 - 01 & 02
413 - 03 - 01 to 09
Electrical separation – one item only
471 - 12 - 01
413 - 06 - 01 to 09
Class II equipment or equivalent
471 - 09 - 01 & 02
413 - 03 - 01 to 09
Non-conducting locations
471 - 10 - 01
413 - 04 - 01 to 07
Earth-free local equipotential bonding
471 - 11 - 01
413 - 04 - 01 to 07
547 - 03 - 01 to 05
Electrical separation – more than one item
412 - 12 - 01
413 - 06 - 01 to 05

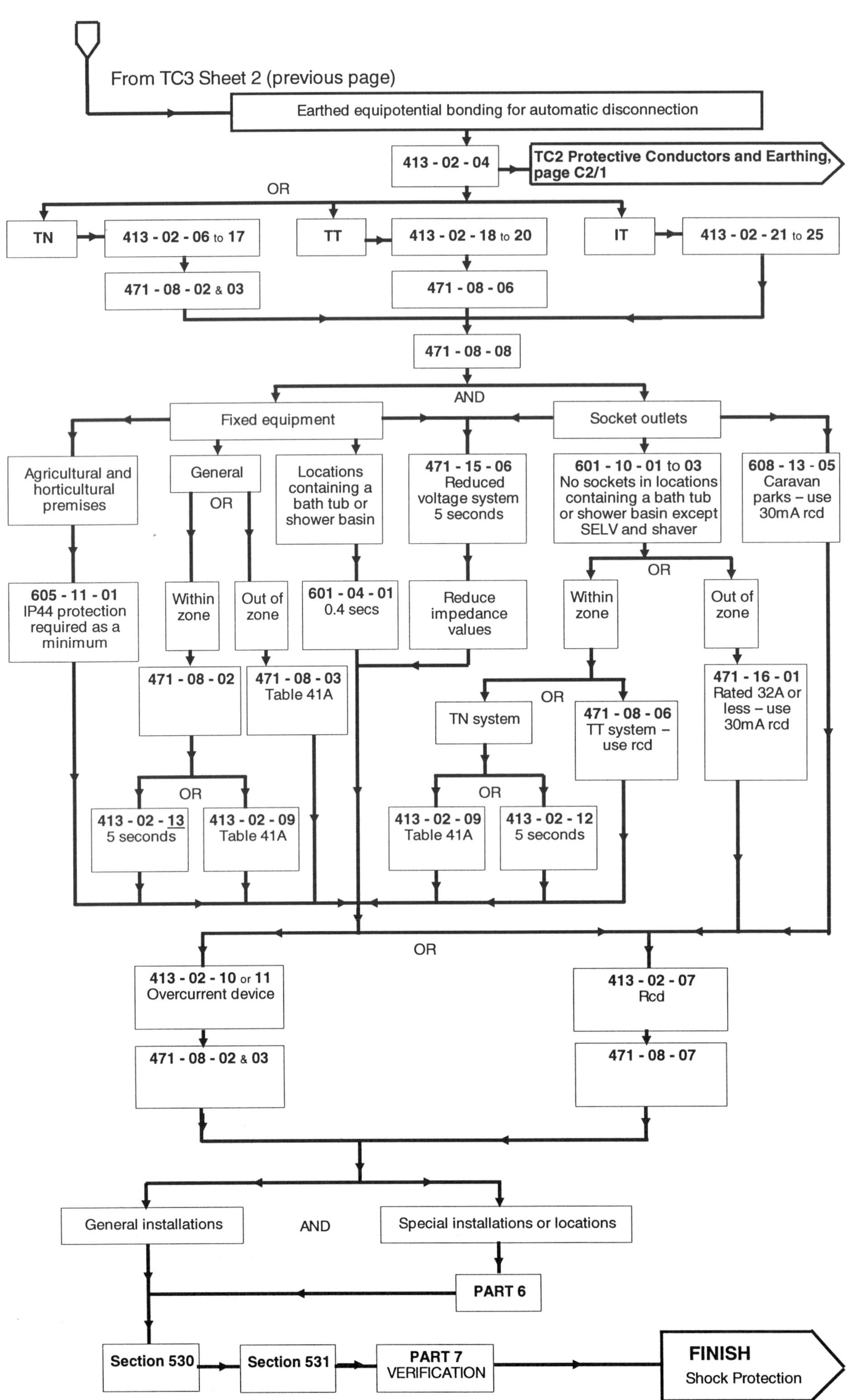

From TC3 Sheet 2 (previous page)
Earthed equipotential bonding for automatic disconnection
413 - 02 - 04
TC2 Protective Conductors and Earthing, page C2/1
OR
TN
413 - 02 - 06 to 17
TT
413 - 02 - 18 to 20
IT
413 - 02 - 21 to 25
471 - 08 - 02 & 03
471 - 08 - 06
471 - 08 - 08
AND
Fixed equipment
Socket outlets
Agricultural and horticultural premises
General
OR
Locations containing a bath tub or shower basin
471 - 15 - 06 Reduced voltage system 5 seconds
601 - 10 - 01 to 03 No sockets in locations containing a bath tub or shower basin except SELV and shaver
608 - 13 - 05 Caravan parks – use 30mA rcd
605 - 11 - 01 IP44 protection required as a minimum
Within zone
Out of zone
601 - 04 - 01 0.4 secs
Reduce impedance values
OR
Within zone
Out of zone
471 - 08 - 02
471 - 08 - 03 Table 41A
TN system
OR
471 - 08 - 06 TT system – use rcd
471 - 16 - 01 Rated 32A or less – use 30mA rcd
OR
413 - 02 - 13 5 seconds
413 - 02 - 09 Table 41A
OR
413 - 02 - 09 Table 41A
413 - 02 - 12 5 seconds
OR
413 - 02 - 10 or 11 Overcurrent device
413 - 02 - 07 Rcd
471 - 08 - 02 & 03
471 - 08 - 07
General installations
AND
Special installations or locations
PART 6
Section 530
Section 531
PART 7 VERIFICATION
FINISH Shock Protection

Combined Protection against Both Direct and Indirect Contact

The scope available to the designer for taking advantage of combined protection is extremely limited, being confined to circuits operating at safety extra-low voltage or a circuit for which the supply is derived from a source having special characteristics. Functional extra-low voltage must not be used alone (411 - 01 - 01) but may be used in combination with other protective measures. These circuits are intended for use in areas of high hazard, e.g. where persons are in damp situations or in close contact with earthed surroundings, such as whilst working inside boilers and pressure vessels (see also Section 606, Restrictive Conductive Locations). It is likely to be necessary to provide combined protection against direct and indirect contact where lighting is provided by handlamps. Such lamps in garages should be shielded against breakage and possible subsequent ignition of fuel.

SELV

The characteristics of a safety extra-low voltage system are that the voltage shall not exceed 50V rms a.c. or 120V ripple-free d.c., precautions having been taken to ensure that the voltage cannot rise above that limit, and that no conductors in the circuit may be connected either deliberately or accidentally with earth, and guarding against accidental short-circuit of high current/low voltage conductors will be necessary to avoid burns or other injuries.

If the voltage of the SELV system is maintained at 25V a.c. (60V d.c.) or less, it is not necessary to provide protection against direct contact for the live parts, i.e. bare conductors are permitted (insulation or physical separation will be required, of course, for functional reasons). Above 25V a.c. (60V d.c.) and up to 50V a.c. (120 d.c.) it is necessary to provide insulation or IP2X barriers or enclosures. Protection against overcurrent is however still required, and guarding against accidental short-circuiting of high current low voltage will be necessary to avoid burns and other injuries.

All of the above voltage limits must be reconsidered and reduced if the location of the installation is such that presence of dampness, water or excessive temperature is to be expected, as in these conditions the resistance of the human body is significantly lowered and the likelihood of dangerous shock increased. These special installations or locations are specified in Part 6.

To ensure that SELV conductors are not earthed accidentally, careful design and installation techniques are required to ensure that separation is effective and maintained. Physical separation is the preferred method, and this implies a totally separate installation with no exposed-conductive-parts connected to earth, to other conductors or metalwork. Thus, it is unlikely that a metal conduit or trunking installation would satisfy this requirement because the possibility of a fixing screw coming into contact with earthed metal cannot be discounted, and neither must the problem of a steel-framed building be ignored.

SELV installations should, therefore, be carried out in insulated conduit or trunking or non-metallic sheathed cables without protective conductors.

SELV plugs and socket-outlets must be of an exclusive pattern, throughout the premises involved. The plugs must not be able to enter (even partially) any other socket-outlet, and the socket-outlets must not accept any other plug or contain provision for connection of a protective conductor.

Functional Extra-low voltage systems

In some installations, extra-low voltage may be selected for operational reasons, but it may not be possible to satisfy all the requirements for SELV circuits. These systems are termed PELV (471 - 14 - 02) when all the requirements of SELV are met <u>except that the system is not electrically separated from earth</u> and 'Functional extra low voltage' where other exceptions to SELV are required, and may, for example, occur where metallic conduit is employed for protection against mechanical damage where Class I equipment is used. In addition, certain components of an extra-low voltage circuit, such as relays, may not have the degree of separation required between contacts carrying extra-low voltage and those contacts or parts connected to a higher voltage circuit; the degree of separation necessary being equivalent to that of a safety isolating transformer to BS 3535. (See the requirements of 471 - 14 - 01 to 06.)

If the system proposed complies with all the requirements for safety extra-low voltage, except that live or exposed-conductive-parts are connected to earth or to the protective conductors of other systems (e.g. one conductor connected to earth or a metallic conduit system) then no additional protection against indirect contact is required, but enclosures to at least IP2X (BS 4590) must be provided, or insulation of live parts must be subjected to an a.c. test voltage of 500V rms for one minute without failure or deterioration.

If there are other features of the installation which do not comply with the safety extra-low voltage requirements, then further protection against both direct contact and indirect contact is required by insulation and/or barriers or enclosures.

In such cases, barriers or enclosures must be provided to prevent unintentional contact, and these are specified in Regulations 412 - 03 - 01 to 04.

Insulation must be suitable for the minimum test voltage of the primary circuit and where accessible, it must be reinforced during erection to withstand a test voltage of 1500V rms for one minute. Exposed-conductive-parts must be connected to the protective conductor of the primary circuit. It follows that such a circuit, although operating at extra-low voltage, must comply with the requirements for a low voltage installation.

If socket-outlets are installed in extra-low voltage circuits other than SELV, they must be of an exclusive pattern different from those in SELV circuits within the same premises, and may contain provision for the connection of a protective conductor.

If any luminaire support coupler has a contact for a protective conductor, it must not be used on a SELV circuit.

NOTE: SELV circuits will only be acceptable under the Electricity At Work Regulations, 1989, provided that all portable equipment is of Class II or Class III insulated construction.

Limitation of Discharge Energy

The Regulations recognize that protection against shock can be afforded if the source of energy is controlled or designed so that the current which can pass through the body of a person or livestock is lower than the shock current, i.e. that current likely to cause danger. There is no limit imposed on the open circuit voltage.

This method has limited application, the best known of which is an electric fence controller. It must be ensured that the source of energy complies with the relevant British Standard for such equipment. It should not be assumed, however, that since the energy discharge itself should not cause danger, there will be negligible risk – e.g. BS EN 61011 requires for electric fence controllers, a peak voltage between 1.5 and 10 kV and a peak current between 150 mA and 10 A. The time pulse is, however, limited to 0.1 seconds, and the current may only exceed 300 mA for 1.5 milliseconds.

Circuits employing limitation of discharge energy as a means of protection against shock must be separated from other circuits as required by Regulations 411 - 02 - 05(i), 05(ii) and 11 for safety-low voltage circuits.

Protection against Direct Contact

The measures which may be adopted to prevent direct contact are one or more of the following:

(i) Protection by insulation of live parts (Regulation 412 - 02 - 01)

(ii) Protection by barriers or enclosures (Regulations 412 - 03 - 01 to 04).

(iii) Protection by obstacles (Regulations 412 - 04 - 01 and 02)

(iv) Protection by placing out of reach (Regulations 412 - 05 - 01 to 04)

It is extremely unlikely that any complete installation encountered in ordinary circumstances will rely on only one of the above measures. It is probable that at least two, protection by insulation and by barriers or enclosures will be used most often, depending on the particular part of the installation being considered.

Protection by insulation of live parts includes insulated cables. Enclosures may be conduit, metal boxes, or insulated boxes where it is only possible to gain access by use of a key or tool such as a screwdriver.

In certain circumstances, e.g. the contacts within a distribution fuse-board, it may not be practical to insulate or enclose live parts. In these circumstances, barriers (e.g. resin bonded board) may be used to prevent unintentional contact; but by definition, will not, and in this example, must not, prevent deliberate action.

It is important to understand that where two methods are used together, one of the methods must meet the requirements fully. For example, if a fuse-board has insulated busbars and connections, the enclosure may be opened without the use of a key or tool. However, it is essential that all live parts within the enclosure be fully shrouded. In other words, the partial application of two different measures of protection cannot be combined to claim compliance with the degree of protection required.

It is important to note the requirements of 471 - 05 - 02. These qualify Regulation 412 - 03 - 01 and allow openings greater than those permitted IP2X, in order to permit items where those requirements are impracticable. The obvious example given is lampholders, which are currently exempted from the Electrical Equipment (Safety) Regulations.

An example of the use of obstacles to prevent direct contact is a handrail in front of an open-type switchboard, but the use of obstacles is applicable where only skilled or instructed persons, under direct supervision, have access (Regulation 471 - 06 - 01). However, certain conditions in Part 6 limit the use of this method and means not only objects in regular use but occasional use such as metal ladders must be taken into account.

Placing out of reach is a method of protection with somewhat restricted application to locations accessible only to skilled or instructed persons. Where bulky or long conducting objects may be handled, it is inadvisable to rely upon this method.

Protection against Indirect Contact

Five basic protective measures for protection against indirect contact (see definition) are specified:

(i) Earthed equipotential bonding and automatic disconnection of supply (Regulations 413 - 02 - 02 to 28 and 471 - 08 - 01 to 471 - 08 - 08.

(ii) Use of Class II equipment or equivalent insulation (Regulation 413 - 03 - 01 to 09 and 471 - 09 - 01 to 471 - 09 - 04).

(iii) Non-conducting location (Regulations 413 - 04 - 02 to 07 and 471 - 10 - 01).

(iv) Earth-free local equipotential bonding (Regulations 413 - 05 - 01 to 04 and 471 - 11 - 01).

(v) Electrical separation (Regulations 413 - 06 - 01 to 05 and 471 - 12 - 01).

The principles underlying the first method are those which have been established practice in the U.K. over many years, and earthed equipotential bonding and automatic disconnection of supply remains the preferred method of protection against indirect contact.

The other four deal with special installations likely to be under the control of trained and experienced persons only. These installations are unlikely to be encountered in normal practice, and they are dealt with later.

Earthed Equipotential Bonding and Automatic Disconnection

The purpose of the bonding of all extraneous-conductive-parts is to create an equipotential zone within which all voltages between exposed conductive parts and extraneous conductive parts are minimized during earth fault conditions. See Notes 1 & 2 (pages C2/1 and C2/5) of this Handbook for an explanation of the necessity or otherwise to bond extraneous-conductive-parts.

The earthing arrangements for the installation, and the impedance of the protective conductors must be co-ordinated with the operating characteristics of the protective device for automatic disconnection of the supply, so that under earth fault conditions the prospective shock voltages within the equipotential zone are limited in magnitude and time duration. The achievement of this co-ordination by using, for example, the limiting values of impedances contained in Tables 41B, 41D Regulations 413 - 02 - 10 to 14 is dealt with in the notes accompanying Topic Chart 2 - "Protective Conductors and Earthing".

The designer and installer must ensure that the earth fault loop impedance is sufficiently low for the disconnection to occur within 0.4 seconds for final circuits supplying socket-outlets, and 5 seconds for final circuits supplying only fixed appliances. The reason for the difference is the assumption that socket outlets are normally used to supply portable and hand-held appliances, so that there is a greater risk of a person being in contact with the appliance at the time of the fault.

It should also be appreciated that the voltage difference between any part of the equipotential zone and "true" earth may be sufficient to cause a dangerous shock current.

It is important to note that the application of the limiting values of earth fault loop impedance contained in Regulations 413 - 02 - 10 and 11and limiting impedances in 413 - 02 - 12 are only applicable to circuits where all exposed-conductive-parts of the equipment concerned, and any extraneous-conductive-parts, are within the main equipotential zone.

Where an installation incorporates socket outlets rated at 32A or less, which may reasonably be expected to supply portable equipment for use outdoors, Regulation 471 - 16 - 01 requires a residual current device (30mA or less operating current and an operating time of not more than 40ms at a residual current of 150mA) in addition to the disconnection time of 0.4 seconds required by 471 - 08 - 02 and Table 41A. For supplies to fixed equipment, a disconnection time of 0.4 seconds or less is required, and this may be achieved by use of an overcurrent protective device or an rcd.

Examples of the application of these requirements are (i) in domestic situations where socket outlets are provided for garden tools, or socket outlets adjacent to openable windows from which garden tools etc. may be fed. Industrial situations where there are socket outlets for the connection of wire brushes, drills, etc., for external use; and (ii) in an industrial situation, such as an outside hoist where the control device may be used inside or outside the zone.

In all installations connected to TT systems, every socket-outlet circuit must be protected by a residual current device, although not necessarily individually.

Residual current devices of the specified operating current and time are only to be considered as additional protection against direct contact. At least one of the measures specified for protection against direct contact must also be used. Devices having the prescribed characteristics may provide additional protection in situations such as domestic gardens where flexible cords for mowers, hedge trimmers, etc., are liable to damage.

Reference should be made to BS 4293 for further information on residual current operated circuit breakers.

Class II Equipment (BS 2754) or Equivalent Insulation

This method relies on the provision of double or reinforced insulation, to present a dangerous voltage appearing on exposed-conductive-parts through the failure of basic insulation. Class II (and for that matter Class I) is a term which applies to appliances.

Apart from Class II fixed appliances forming part of an installation, some accessories and other equipment may be of a construction equivalent to Class II. Where, for example, insulated plate-switches are installed, a circuit protective conductor must be provided since there is the possibility that one might later be replaced by a metal accessory.

Installations where it can be safely assumed that only Class II equipment will ever be installed or used will occur very rarely, and only in special circumstances where they are under effective supervision in normal use.

Class II equipment should not be earthed either deliberately or fortuitously because this could impair the protection against indirect contact afforded by the equipment specification.

Moreover, if it is desired to mount Class II equipment on earthed metalwork, e.g. a Class II luminaire mounted on a metal conduit box, it must be ensured that basic insulation is not the only protection between live parts of the installation and the exposed metalwork of the Class II equipment. If the mounting face of the Class II equipment is metallic, this requirement implies the provision of an insulating separator between the equipment and the metal box.

The explicit requirements of these Regulations for the provision and testing of insulating enclosures, particularly if they are site constructed, require special consideration and experience (see Inspection and Testing on page C10/14).

Note on Classification of Means of Protection Against Electric Shock

Full information on the various classifications is given in BS 2754 to which reference should be made, but a summary of the methods is given below. This BS is based on I.E.C. Report No.536 and provides for four classifications which do not indicate the degree of protection, but only the method of achieving it.

Class 0, in which protection is provided only by basic insulation; no provision is made for the earthing of accessible conductive parts. Such an arrangement would only be acceptable in an earth-free situation and since this cannot be continuously guaranteed, Class 0 is **not** accepted in the United Kingdom, and there are statutory prohibitions against its use.

Class I relies on basic insulation, and the earthing of all accessible conductive parts. It follows that to comply with the BS, the flexible cord of Class I equipment must include a protective conductor, connecting the accessible conductive parts to the circuit protective conductor.

Class II equipment does not rely solely on basic insulation, but has double or reinforced insulation. No provision is made for the connection of a protective conductor to the accessible conductive parts, (if any).

Class III equipment relies for protection on the use of Safety Extra-Low Voltage (SELV) supplies, and no voltages higher than ELV may be generated within the equipment. (ELV is a.c. of not more than 50V, or d.c. not exceeding 120V and ripple-free. See the appropriate sections of this Handbook for further information.)

N.B.

Because of the Electricity At Work Regulations, 1989 requirement to earth the external metalwork of all portable tools and equipment, only Class II, or Class III, insulated equipment may be used in industrial situations on SELV supplies.

Protection by Non-Conducting Location

In a non-conducting location, reliance is placed on the provision of insulating walls and floors and the separation by more than two metres of two exposed-conductive-parts, or an exposed conductive part and any extraneous conductive part.

This method is intended to prevent simultaneous contact with exposed-conductive-parts and/or extraneous-conductive-parts which may be at different potentials following failure of basic insulation.

The non-conducting location must have insulating floors and walls, and the spacing between conducting parts must be at least 2 metres, although this spacing may be reduced to 1.25 metres outside the zone of arm's reach (see Regulation 412 - 05 - 02). Insulating obstacles may be installed to achieve the effective spacing. As an alternative to spacing, extraneous-conductive-parts may be insulated. Protective conductors must not be introduced into the non-conducting location and it must be ensured that any potential on extraneous-conductive-parts within the location cannot be transmitted outside the location. Additionally, socket outlets must not incorporate an earthing contact.

This method should be used only in special situations where specified by a suitably qualified electrical engineer. For example, it may be adopted for a test room under supervision adequate to ensure that no change occurs in the design conditions, and where trained persons only are permitted to work.

Special consideration must be given to the design, installation and testing of non-conducting locations (see Inspection and Testing on page C10/14).

Earth-free Local Equipotential Bonding

In such locations, all exposed conductive parts and extraneous conductive parts which might be accessible at any one time shall be bonded together by an equipotential conductor which must itself not be in contact with earth.

The difficulties inherent in applying the method are obvious. Metal window frames and ceiling suspension channels, if exposed, are among the items to be bonded.

It must be ensured that, in the event of failure of basic insulation, all exposed metalwork in the location will be at the same potential, and conditions for shock will not exist. Thus, the use of this method requires special consideration, and again is only likely to be applied in an industrial testing situation.

Protection by Electrical Separation

This method may be applied to single items of equipment, such as a shaver unit complying with BS 3052 or, where specified by a suitably qualified electrical engineer, to several items of equipment, in which case the additional rules in Regulation 413 - 06 - 05 must be observed, and the separated installation must be under effective supervision.

The degree of electrical separation must be not less than that between the input and output of a safety isolating transformer complying with BS 3535. The source of supply for the separated circuit may be for example:

(i) A safety isolating transformer complying with BS 3535.

(ii) Motor generator having separate windings.

Residual Current Devices (rcds)

A residual current device is defined in the Regulations as "a mechanical switching device or association of devices, intended to cause the opening of the contacts when the residual current attains a given value under specified conditions". Thus an rcd may be a self-contained piece of equipment, i.e. a residual current circuit breaker (rccb), or it may be a detection/control unit operating a separate circuit breaker, which may be convenient when larger loads require to be protected. An rcd may also incorporate over-current tripping facilities. There are several ways of achieving residual current (previously known as current operated earth leakage) protection as required by the Definition (although the results are the same). It is necessary to know and differentiate between the various types both to secure the desired performance and to avoid damage during testing. It is also necessary to know the parameters against which rcds are designed in order to appreciate what may be expected of them.

All rcds depend upon the action of a current transformer to compare the phase and neutral currents (for single phase), and phases and neutral, if any, for three phase devices. A residual current i.e. one caused by current being conducted by other than the circuit in question, is detected and used to operate the device, opening the associated live conductors. Since rcds must operate on small currents, the initiating power is limited and in some devices is amplified to obtain adequate tripping force. This may be done by self contained electronic devices, powered from the supply passing through the device, or a sensitive relay may operate the trip by an auxiliary supply (see 531 - 02 - 06). It is not always apparent when an rcd incorporates electronic components, and may therefore be sensitive to damage during installation testing. Manufacturers' literature does not always make this clear, and enquiries may have to be made to ascertain whether 500V d.c. testing is permissible. Other rcds depend upon magnetic operation in various forms, and must therefore be situated where they are not subject to stray magnetic fields. (531 - 02 - 07).

It is important to recognize the tolerances in rcd operation laid down in BS 4293. The range is from 50% to 100% of the rated tripping current, e.g. a 30mA device must operate between 15 and 30mA, so that a device which may appear over-sensitive could well meet the requirements of the BS. An rcd to BS 4293 has a maximum breaking capacity of 500A, and a through fault current withstand of around 3kA.

Another point worth noting is that the test push provided on rcds does not necessarily apply only the rated tripping current to confirm operation. In accordance with the BS the test current may be up to 2.5 times the rated tripping current, one reason being that since the test resistor must permit operation on supply voltages less than 230V, when used on 230V the test current must be higher than nominal. Operation of the test push does not prove that the device trips at its rated current, neither does it check the protective conductors or any earth electrode of the installation; it only indicates that the device itself operates with currents of the correct order of magnitude.

Although rcds are designed to give protection against direct and indirect contact, total reliance must not be placed upon them and other recognized methods of protection must still be provided (see Regulations 412 - 01 - 01, 412 - 06 - 01 and 02, and 413 - 01 - 01).

It should be noted that rcds cannot be relied upon to provide discrimination according to their tripping current ratings. This is due partly to the tolerances already mentioned and partly to the fact that fault currents may be of any magnitude, so that if a 30mA rcd is supplied through a 100mA device, they will trip simultaneously with 100mA, and maybe with 50mA; although the smaller device would clear 30mA on its own. If discrimination has to be provided, then it may be done by a time delay on the 'upstream' rcd with the larger current rating. Reference should be made in this connection to the paragraph on 'Minimizing Danger and Inconvenience' on page C1/3. Regulation 413 - 02 - 19 recommends that the whole of an installation connected to a TT supply be protected by an rcd, and 471- 08 - 06 requires all socket outlets on such a system to be protected by an rcd. To satisfy 314 - 01 - 01 and 02 without providing a main isolating switch connected to two rcds, virtually requires the use of a split-load consumer unit designed for just this purpose or one incorporating a time delayed 100mA 'main' rcd (providing 'Isolating' contact clearances), and with a subsidiary 30mA rcd for the socket outlets.

There are three specific requirements in the Regulations for the provision of rcds and these are:

(1) 471- 08 - 06 requires it for all socket outlet circuits connected to a TT system.

(2) 608 - 13 - 05 requires a 30mA device for supplies to mobile caravans.

(3) 471 - 16 - 01 requires a 30mA device for any socket outlet with a rating of 32A or less likely to be used to supply equipment outside the equipotential bonded area.

N.B. A situation of which it is as well to be aware, can occur on installations where a large proportion of the load consists of equipment with solid-state controls, e.g. dimmers, speed controls, and electronic apparatus. These can all give rise to harmonics which distort the basic sine waveform of the supply, and may give rise to something more resembling a pulsating d.c. waveform. If severe enough this can result in tripping only at values higher than the rcd rating, or possibly failure to trip. rcds are available which are designed to operate satisfactorily with such loads. They are more expensive, but the possible need for them should be kept in mind.

Figure C7

Sockets & Fixed Equipment Outside Main Equipotential Zone

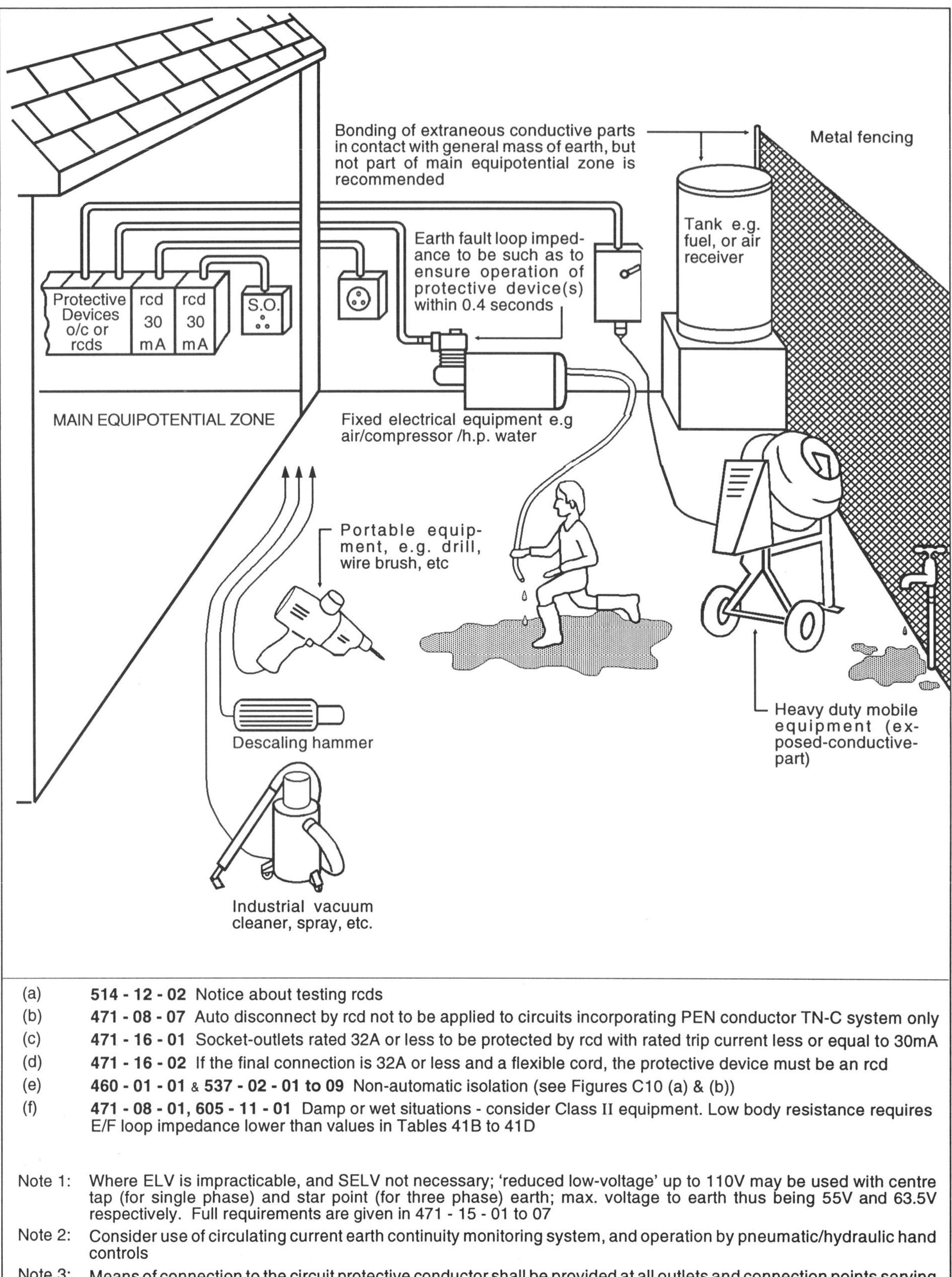

(a) **514 - 12 - 02** Notice about testing rcds

(b) **471 - 08 - 07** Auto disconnect by rcd not to be applied to circuits incorporating PEN conductor TN-C system only

(c) **471 - 16 - 01** Socket-outlets rated 32A or less to be protected by rcd with rated trip current less or equal to 30mA

(d) **471 - 16 - 02** If the final connection is 32A or less and a flexible cord, the protective device must be an rcd

(e) **460 - 01 - 01** & **537 - 02 - 01 to 09** Non-automatic isolation (see Figures C10 (a) & (b))

(f) **471 - 08 - 01, 605 - 11 - 01** Damp or wet situations - consider Class II equipment. Low body resistance requires E/F loop impedance lower than values in Tables 41B to 41D

Note 1: Where ELV is impracticable, and SELV not necessary; 'reduced low-voltage' up to 110V may be used with centre tap (for single phase) and star point (for three phase) earth; max. voltage to earth thus being 55V and 63.5V respectively. Full requirements are given in 471 - 15 - 01 to 07

Note 2: Consider use of circulating current earth continuity monitoring system, and operation by pneumatic/hydraulic hand controls

Note 3: Means of connection to the circuit protective conductor shall be provided at all outlets and connection points serving Class II equipment which may be changed by the user (471 - 09 - 02)

Figure C8

Safety Extra-Low Voltage Sources

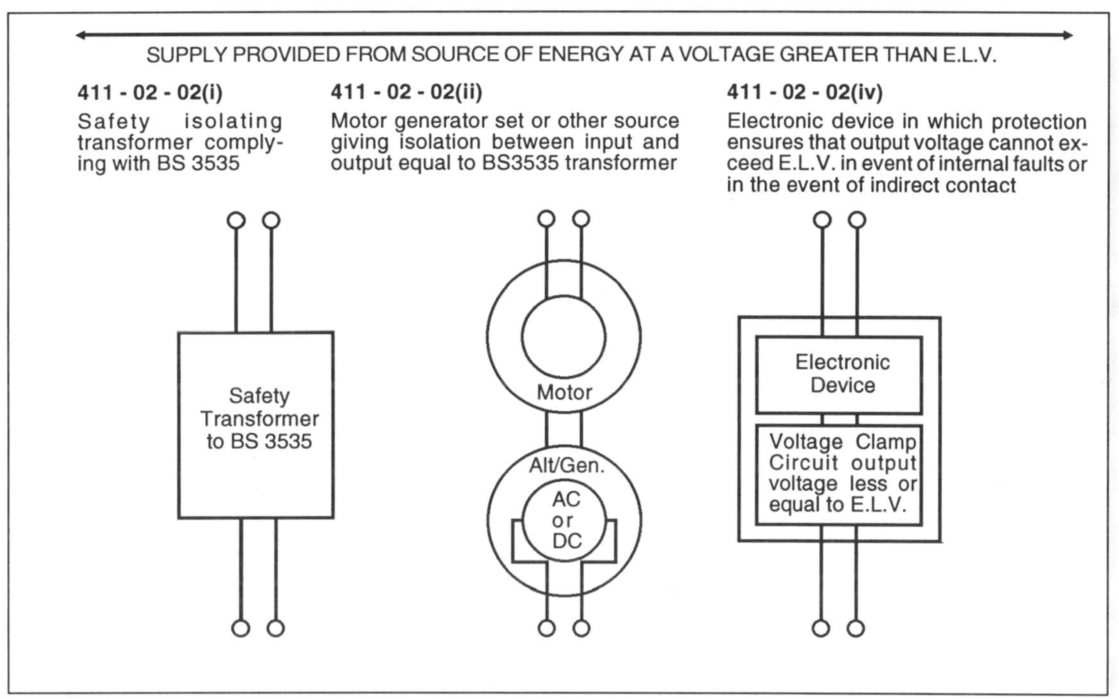

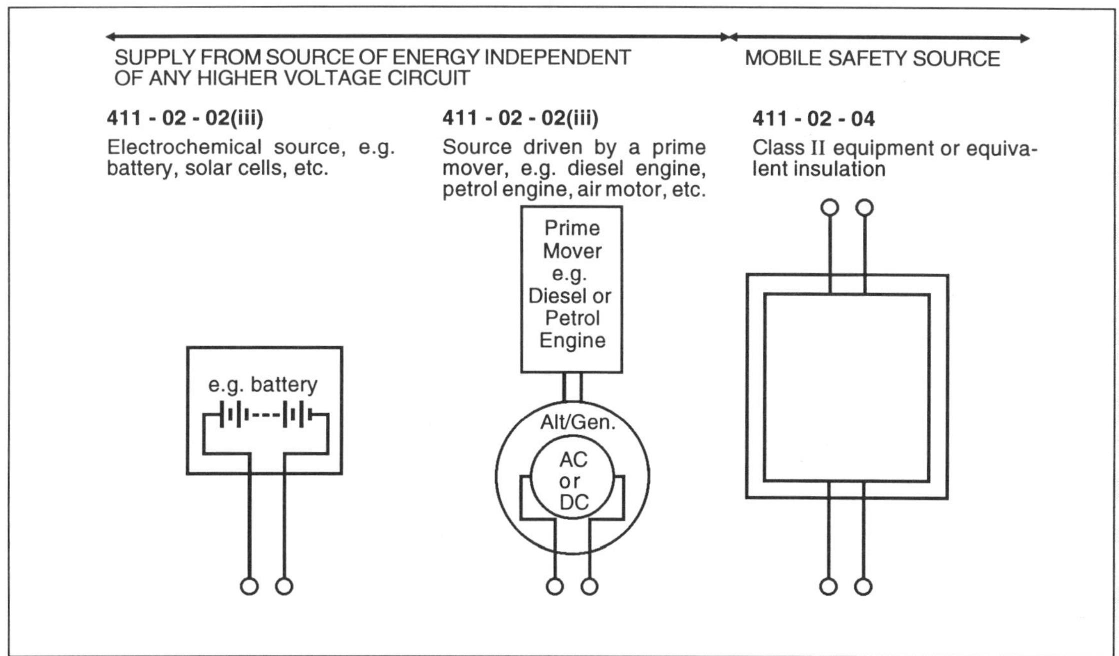

Note:

Safety Extra-Low Voltage supplies may be obtained from the six types of sources shown. The nominal voltage for such supplies must not exceed 50V rms a.c. or 120V ripple free d.c., and this must be reduced to 25V a.c. (60V d.c.) or less, in situations of unusual risk, e.g. damp or confined spaces (604 - 02 - 02). SELV output supplies and circuits must be isolated from all other conductors and exposed-conductive-parts. If voltage is between 25V and 50V a.c. (60V and 120V d.c.) then protection against direct contact must be provided by IP2X or IPXXB enclosures or insulation to withstand 500V a.c. rms test for one minute (411 - 02 - 09)

A bell transformer is only a safety source if it is a Class II safety isolating transformer, double insulated to BS 3535. If it is a Class I or not to BS 3535, then the bell circuit will be anl Extra-low voltage system other than SELV, and must be earthed at one point, any exposed-conductive-parts being connected to the primary circuit protective conductor. Older bell transformers with cores which accept only single strand conductors with sheath removed contravene Regulation 526 - 03 - 03 and are not acceptable.

For details of SELV and Extra low voltage systems other than SELV see page C1/8. Where the source is a 'generator' see the requirements of Section 551 and page D1/6.

Topic Chart 4

AN INFORMATION CHART

Isolation and Switching

Isolator. A mechanical switching device which, in the open position, complies with the requirements specified for isolation. An isolator is otherwise known as a disconnector.

Isolation. A function intended to cut off, for reasons of safety, the supply from all or a discrete section of the installation by separating the installation or section from every source of electrical energy.

Switch. A mechanical device capable of making, carrying and breaking current under normal circuit conditions, which may include specified operating overload conditions, and also of carrying for a specified time currents under specified abnormal circuit conditions such as those of short-circuit. It may also be capable of making, but not breaking, short-circuit currents.

Mechanical Maintenance. Mechanical maintenance is the replacement, refurbishment or cleaning of lamps and non-electrical parts of equipment, plant and machinery.

Emergency Switching. An operation intended to remove as quickly as possible, danger, which may have occured unexpectedly.

Functional Switching. An operation intended to switch 'on' or 'off' or vary the supply of electrical energy to all or part of an installation.

See Part 2 of BS 7671: 1992 for the complete set of Definitions

TOPIC CHART 4 (Information) Isolation and Switching

START

130 - 06 - 01 & **02** Fundamental requirements for isolation and switching

314 - 01 - 01 to **03**
Division of installation into circuits to avoid danger, minimise inconvenience in case of fault, and facilitate safe operation, inspection, testing and maintenance

460 - 01 - 01 and **02**
Requirement for non-automatic (i.e. manually operated) isolation and switching

460 - 01 - 03
Prohibits isolation and switching for PEN and protective conductors

476 - 01 - 01 and **537 - 04 - 06**
Use of devices for more than one function of isolation and switching

Section 530
Common requirements for switchgear

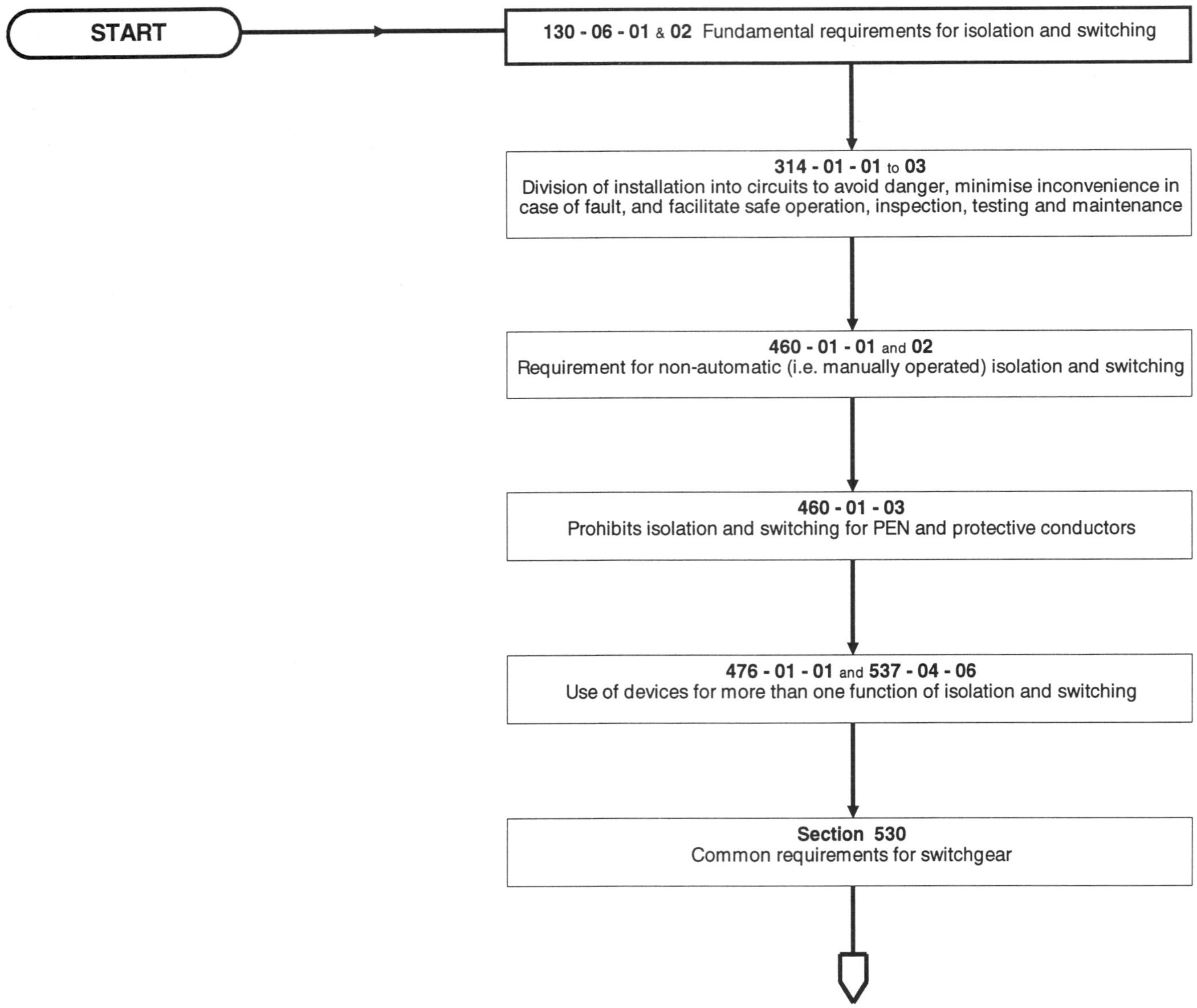

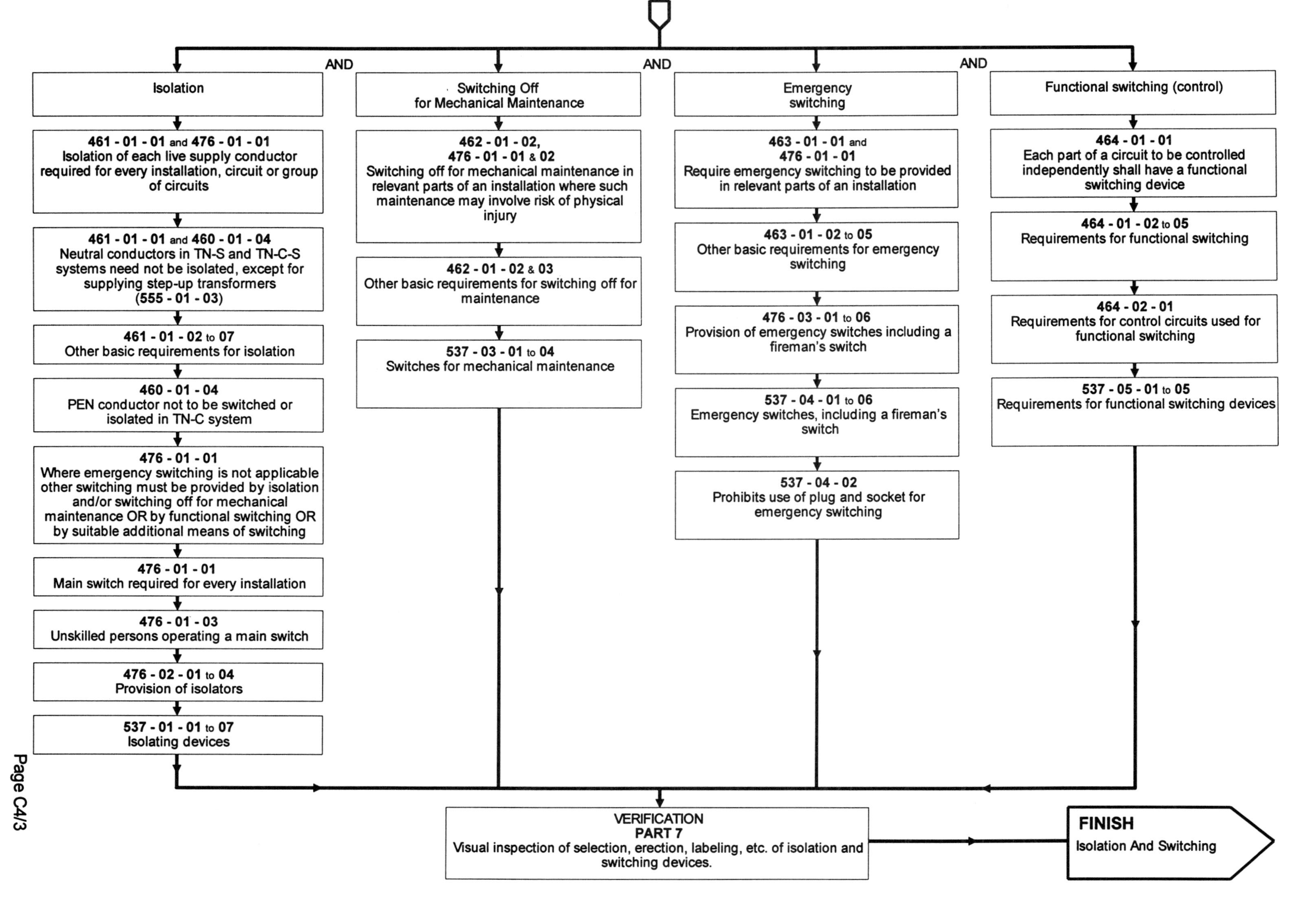

AND
AND
AND
Isolation
461 - 01 - 01 and 476 - 01 - 01
Isolation of each live supply conductor required for every installation, circuit or group of circuits
461 - 01 - 01 and 460 - 01 - 04
Neutral conductors in TN-S and TN-C-S systems need not be isolated, except for supplying step-up transformers (555 - 01 - 03)
461 - 01 - 02 to 07
Other basic requirements for isolation
460 - 01 - 04
PEN conductor not to be switched or isolated in TN-C system
476 - 01 - 01
Where emergency switching is not applicable other switching must be provided by isolation and/or switching off for mechanical maintenance OR by functional switching OR by suitable additional means of switching
476 - 01 - 01
Main switch required for every installation
476 - 01 - 03
Unskilled persons operating a main switch
476 - 02 - 01 to 04
Provision of isolators
537 - 01 - 01 to 07
Isolating devices
Switching Off for Mechanical Maintenance
462 - 01 - 02,
476 - 01 - 01 & 02
Switching off for mechanical maintenance in relevant parts of an installation where such maintenance may involve risk of physical injury
462 - 01 - 02 & 03
Other basic requirements for switching off for maintenance
537 - 03 - 01 to 04
Switches for mechanical maintenance
Emergency switching
463 - 01 - 01 and
476 - 01 - 01
Require emergency switching to be provided in relevant parts of an installation
463 - 01 - 02 to 05
Other basic requirements for emergency switching
476 - 03 - 01 to 06
Provision of emergency switches including a fireman's switch
537 - 04 - 01 to 06
Emergency switches, including a fireman's switch
537 - 04 - 02
Prohibits use of plug and socket for emergency switching
Functional switching (control)
464 - 01 - 01
Each part of a circuit to be controlled independently shall have a functional switching device
464 - 01 - 02 to 05
Requirements for functional switching
464 - 02 - 01
Requirements for control circuits used for functional switching
537 - 05 - 01 to 05
Requirements for functional switching devices
VERIFICATION
PART 7
Visual inspection of selection, erection, labeling, etc. of isolation and switching devices.
FINISH
Isolation And Switching

Isolation and Switching

Every installation must be provided with a main switch and a means of isolation, which may be combined in a single device.

Every circuit or group of circuits must be provided with:

(i) a means of isolation (a disconnector);

(ii) a means of interrupting the supply on load; and

(iii) a means of interrupting the supply in any anticipated fault conditions.

The purpose of isolation is to ensure the safety of skilled and/or instructed persons by disconnection of live parts from the supply.

Means of switching off for mechanical maintenance must be provided for every circuit supplying:

- an electric motor
- electrical heating equipment with exposed-to-touch surfaces
- electromagnetic equipment which might cause mechanical accidents
- luminaires (lamp replacement and cleaning are defined as mechanical maintenance)
- any other electrically energized equipment from which possible mechanical or heat hazards could arise from the use of electrical energy.

Emergency switching is required for every part of an installation where it may be necessary to disconnect from the supply rapidly in order to prevent or remove a hazard.

Motors and their starting equipment, individually or in groups, are required to have a means of isolation. Switching for mechanical maintenance and functional switching must also be provided for every motor circuit, and emergency switching must be provided where any electrically driven machine may give rise to danger. Every appliance or luminaire connected other than by means of a plug and socket-outlet, must be controlled by a switch.

Generally, a device may provide more than one of the required means of isolation and switching required by the Regulations, if it is suitable in every respect for each purpose.

Isolation

Intended primarily for the use and protection of electrically skilled and/or instructed persons.

An isolator is a device for manual operation which has to be capable of:

(a) opening and closing a circuit under no-load conditions; and

(b) carrying the normal circuit current; and

(c) carrying for a specified time, abnormal currents which may occur during an over-current condition (overload or short-circuit).

The requirements for isolating devices are set out in Regulations 537 - 02 - 01 to 08 and may be summarized as follows:

(i) The device must conform to BS EN 60947-3 in respect of creepage and clearance distances. However, BS EN 60947-3 does not yet contain a requirement for contact separation distance, which has still to be agreed but it is anticipated that the distance will be between 3 and 4mm for low voltage equipment.

(ii) There must be clear indication as to whether the device is in the open or closed position. The indicator must only show that the isolator is open and in the safe position when all the contacts have achieved the specified separation. Alternatively, the position of the contacts must be visible.

(iii) The device must not be capable of being reclosed unintentionally. This requirement may be satisfied either by a feature of its design or the method of its installation.

(iv) An off-load isolating device (isolator) must be capable of being secured against inadvertent or unauthorized operation, for example, by padlocking, placing in a lockable space or enclosure, or by interlocking with a load-breaking device.

Both the definition and the requirements admit the use of switch-fuses, isolating switches and circuit-breakers, provided they meet the contact clearance requirement, but it will be readily apparent that these isolation requirements preclude the use of certain devices such as:

(a) Switches and circuit-breakers (mcbs, rcds, etc.) intended for use on a.c. only, which have less than the specified clearances. "Micro-gap" switches are clearly not acceptable for this purpose.

(b) As stated in Regulation 537 - 02 - 03, no semiconductor device such as a "touch-switch" or "photo-electric switch" may be used for isolation.

The following devices are capable of satisfying the requirements for isolation depending on the particular installation requirements:

(1) Isolators (for off-load disconnection only).
(2) Isolating switches (for on-load disconnection and incorporating required isolating distance).
(3) Links, including fuse-links.
(4) Plugs and socket-outlets.
(5) Circuit-breakers, complying with the contact separation requirement.

The means of isolation provided must disconnect all live conductors – except as noted below. A neutral conductor is (by definition) a live conductor and where a link is inserted it must comply with 537 - 02 - 05.

<u>Protective conductors shall not be provided with a means of isolation except:</u>

<u>a.</u> <u>Regulation 476 - 01 - 03 requires a main switch for operation by an unskilled person to interrupt both live conductors of a single phase supply.</u>

<u>b.</u> <u>Regulation 555 - 01 - 03 requires a linked switch to disconnect all live conductors of the supply to a step up transformer.</u>

<u>When the requirements of a. and b. are met on a TN-C or TN-C-S supply system the protective conductor will have a means of interruption but not isolation.</u>

The requirement of Regulation 542 - 04 - 02 for a means of disconnection of the main earthing terminal from the means of earthing may, at first sight, appear to be in conflict with the prohibition of isolation or switching of the protective conductor. This is not so, as the means of disconnection is provided specifically for test purposes when the installation is not in normal use and is under the complete control of electrically-skilled persons. However, when such tests are made, the entire installation associated with the means of earthing must be isolated.

An isolating device which is not required to perform any switching function other than isolation may be an off-load device, in which case it must be capable of being secured against inadvertent or unauthorised operation.

Isolating devices (on-load or off-load) may be installed adjacent to the equipment to be isolated, so that they are under the control of the electrically skilled person working on that equipment. If isolating devices are installed remotely from the equipment to be isolated, provision must be made for securing them against inadvertent reclosure, for example by means of unique keys or non-interchangeable removable handles.

Every isolator must be identified to indicate clearly the circuit with which it is associated.

Switching off for Mechanical Maintenance

– For use and protection of other than electrically skilled persons.

The devices provided for this purpose are solely to ensure that equipment is safe for persons who are not necessarily electrically skilled to work upon, and no access to live parts is involved or indeed permitted.

Examples of mechanical maintenance are the replacement of lamps and cleaning of luminaires and the cleaning or repair of electrically-driven machines. Where there is possible danger, a means of isolation for mechanical maintenance is necessary to provide additional safety precautions. Examples of such situations are high voltage discharge lighting installations, or where it is necessary to enter or work in close proximity to normally moving parts, such as a rotating drum or a conveyor. The device for this purpose may be inserted in the supply conductors or in the control circuits.

It is preferable to adopt the first alternative, taking care that the device:

(a) has a reliable or visible indication of the open or off position;

(b) is so designed or installed as to make a positive action necessary to close it;

(c) is capable of breaking the full load current of the equipment connected to it; and

(d) is conveniently situated, readily identifiable and provided with suitable means to prevent unintentional re-closure, caused by e.g. vibration.

In some circumstances, the device may be installed in the control circuits, but such an arrangement requires special consideration to ensure that the degree of safety achieved is not less than that of cutting off the main supply. In general, this will require supplementary precautions such as the use of padlocks or integral locks on the control circuit device. The hazard involved may be burns or physical injury caused by mechanical movement, and warning labels may be required on other control devices such as motor starters, to warn of the presence of two supplies. In such circumstances, a single act of switching off for mechanical maintenance may not be sufficient.

Emergency Switching

May be operated by anyone, not necessarily skilled.

As the name clearly indicates, devices provided for this purpose are for cutting off the supply as rapidly as possible to remove a foreseeable dangerous condition. The hazard involved may be electrical (i.e. danger from shock or burns) or mechanical (i.e. danger from movement).

It should be noted that the means provided should act as directly as possible on the supply conductors. The clear implication of this is that direct-acting means should be provided, rather than reliance on other electrical circuits.

A device for emergency switching must be capable of cutting off the full load current of the circuit, including stalled motor current where appropriate.

A push-button which actuates the switch mechanism directly is preferred, although stop-lock or stop-latch push-buttons acting as switches to tripping relays are not prohibited, where they are under the control of the person at risk. Release of such operating devices must not re-energise the equipment unless clear warning is given that this will occur.

Where delays are provided on the closing or opening of contactors, e.g. for undervoltage protection, this must not prevent instantaneous action where necessary for emergency switching.

The consequential effects of emergency switching must be taken into account. For instance, a particular process might involve electric heating elements under which articles pass on a conveyor belt. While it is necessary to provide emergency switching to stop the conveyor belt, it is important that the heating elements should be disconnected at the same time to reduce the risk of fire.

Where there is danger from moving machinery, emergency stopping is necessary. In such situations, the operation of an emergency switch must cause all movement to be rapidly arrested. As well as cutting off the supply, adequate braking arrangements may be needed for high inertia machines and drives.

Obviously, it is necessary to ensure that all devices for emergency switching are accessible and readily identifiable. Attention should be given to this most important point, preferably in consultation with those who are to operate the machinery or plant.

Plugs and socket-outlets are not permitted as a means of emergency switching.

Functional Switching (Control)

This shall be provided for each circuit or part of a circuit that requires control independently of other parts of the circuit or installation.

This is the term used to denote the normal switching necessary for control of electrically operated equipment of all kinds that may be operated by anyone, not necessarily skilled.

Devices for switching off for mechanical maintenance, or for emergency switching, can also provide switching or control for the normal operation of equipment. If functional switching is not provided in this manner, it is necessary to provide a switch capable of interrupting the on-load supply for any circuit or appliance.

Plugs and socket outlets may be used for functional switching. Plugs and sockets rated at more than 16A must not be used for switching d.c. circuits.

Semi-conductor devices may only be used for functional switching and control subject to compliance with Section 512 of BS 7671: 1992 regarding operational conditions and external influences. They must not be used for any other switching function.

Combined Devices

Mention was made earlier that it is permissible to combine the functions of isolation, switching off for mechanical maintenance, and emergency switching in one device, provided that the device satisfies all the requirements for each function.

A pole operated fireman's switch in a discharge lighting installation, is an example of a device which may perform the functions of isolation and emergency switching. Depending on its situation in relation to the lighting installation, it may also be suitable for the purpose of mechanical maintenance. However, the requirements for this purpose would not be satisfied if the lighting installation was not reasonably near the switch, and the switch did not have provision for locking-off.

Figure C9(a)

Isolation and Switching

(for identification and notices see Regulations Section 514, and for accessibility see 513)

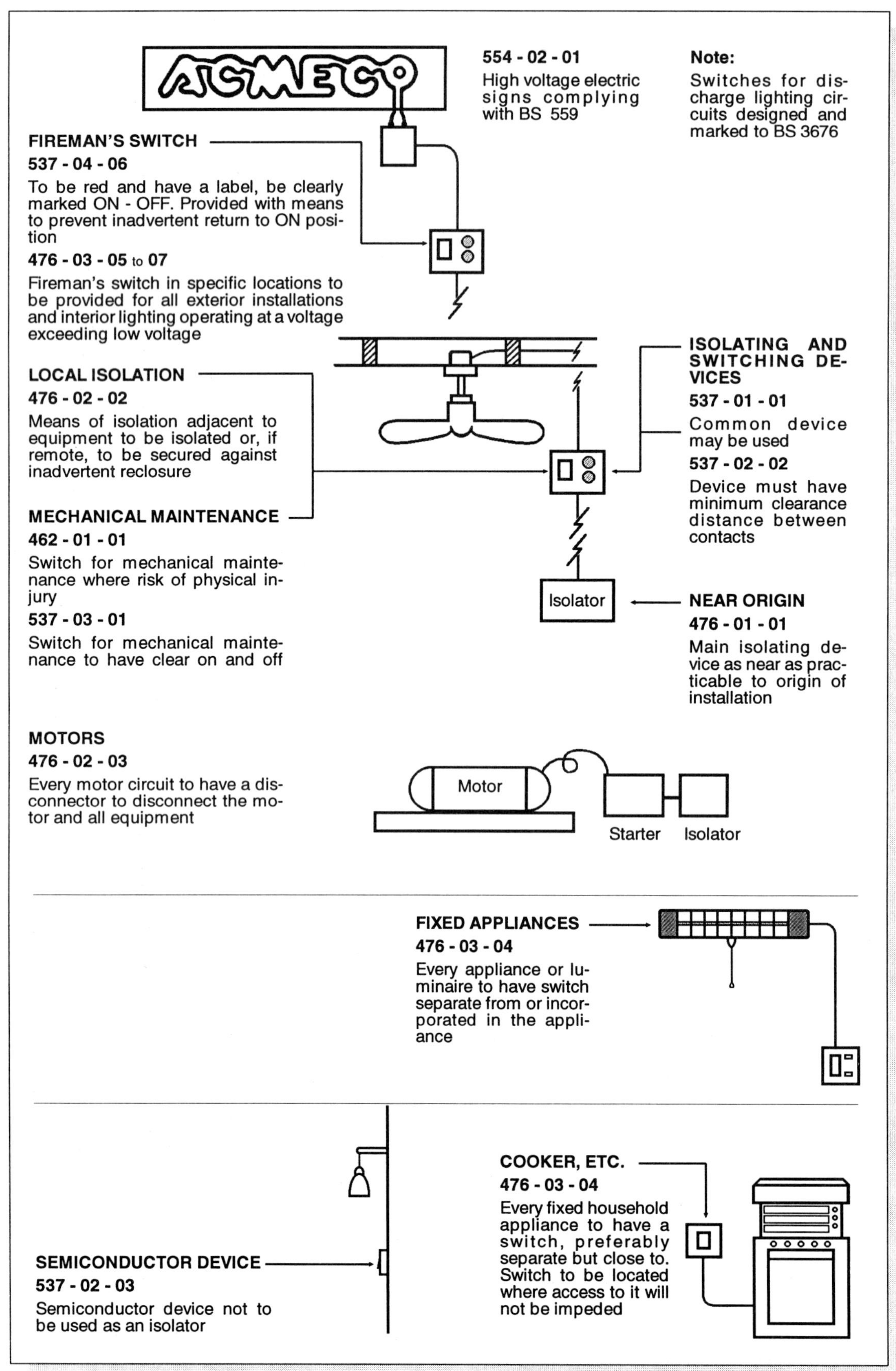

Figure C9(b)

Isolation and Switching (continued)

(for identification and notices see Regulations Section 514, and for accessibility see Section 513)

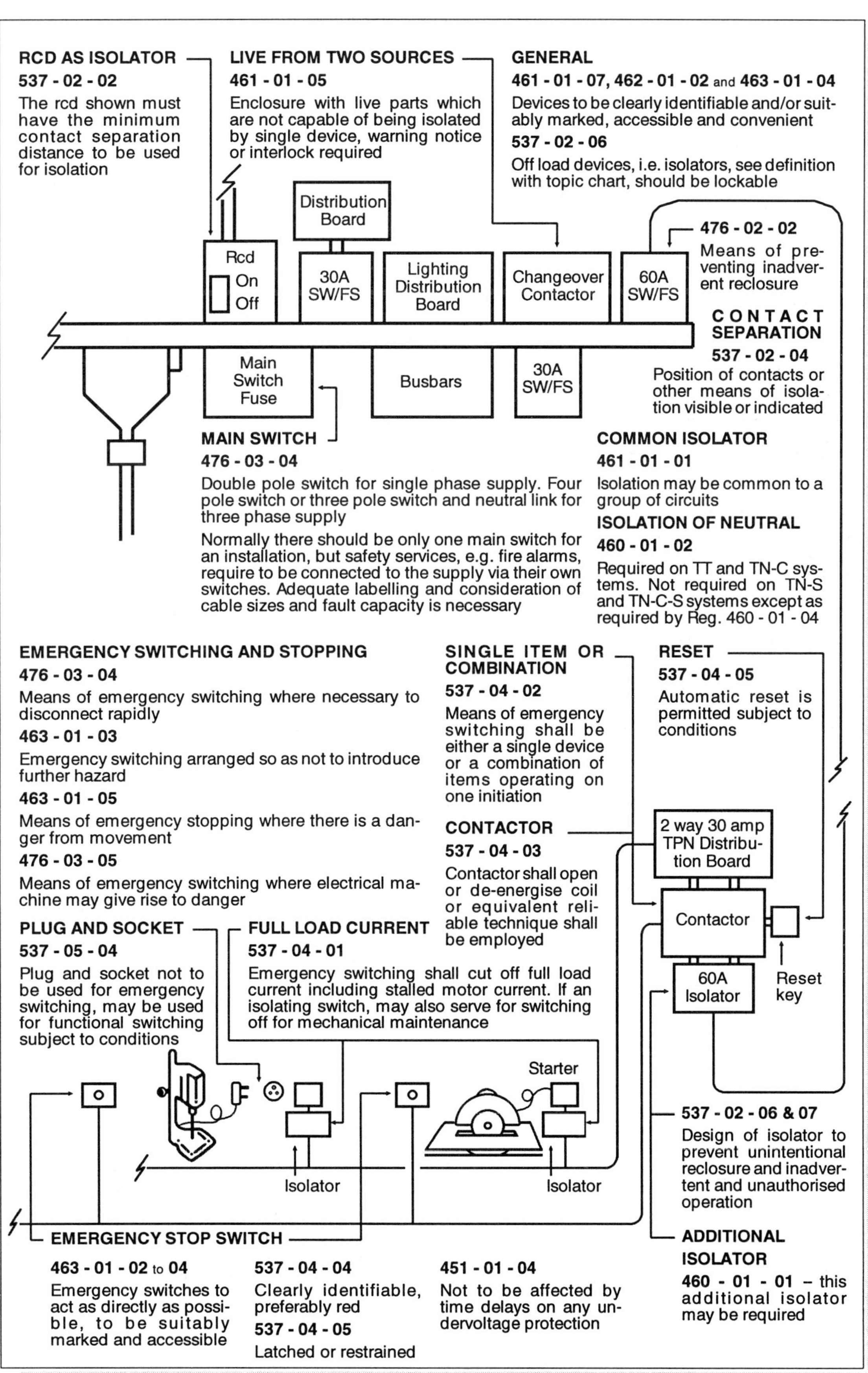

Topic Chart 5

A DECISION CHART

Protection against Overcurrent

Overcurrent is a current exceeding the rated value for a current-carrying part. It may be either an overload current or a short-circuit current. For conductors the rated value is the current carrying capacity

Overload Current is an overcurrent occurring in a circuit which is electrically sound, due to loading in excess of the design current of the circuit.

Short-circuit Current is an overcurrent resulting from a fault of negligible impedance between live conductors having a difference in potential under normal operating conditions.

Earth Fault Current is a fault current which flows to earth. This may be of a similar magnitude to a short circuit current if the fault between a live conductor and earth is of negligible or low impedance (for Protective Measures, see Section 413).

Fault is a circuit conditon in which current flows through an abnormal or unintended path. This may result from an insulation failure or the bridging of insulation. Conventionally the impedance between live conductors or between live conductors and exposed- or extraneous-conductive-parts at the fault position is considered negligible.

Fault Current is a current resulting from a fault.

Prospective Fault Current is the value of overcurrent at a given point in a circuit resulting from a fault of negligible impedance between live conductors having a difference of potential under normal operating conditions, or between a live conductor and an exposed-conductive-part.

In the Topic Chart, conditions to be satisfied include the following quantities:-

I_b design current of circuit
I_n nominal current or current setting of protective device
I_z current-carrying capacity of a cable (when installed)
I_2 current causing effective operation of the overload protective device
I fault current

See Part 2 of BS 7671: 1992 for the complete set of Definitions

TOPIC CHART 5 (Decision) Protection against Overcurrent

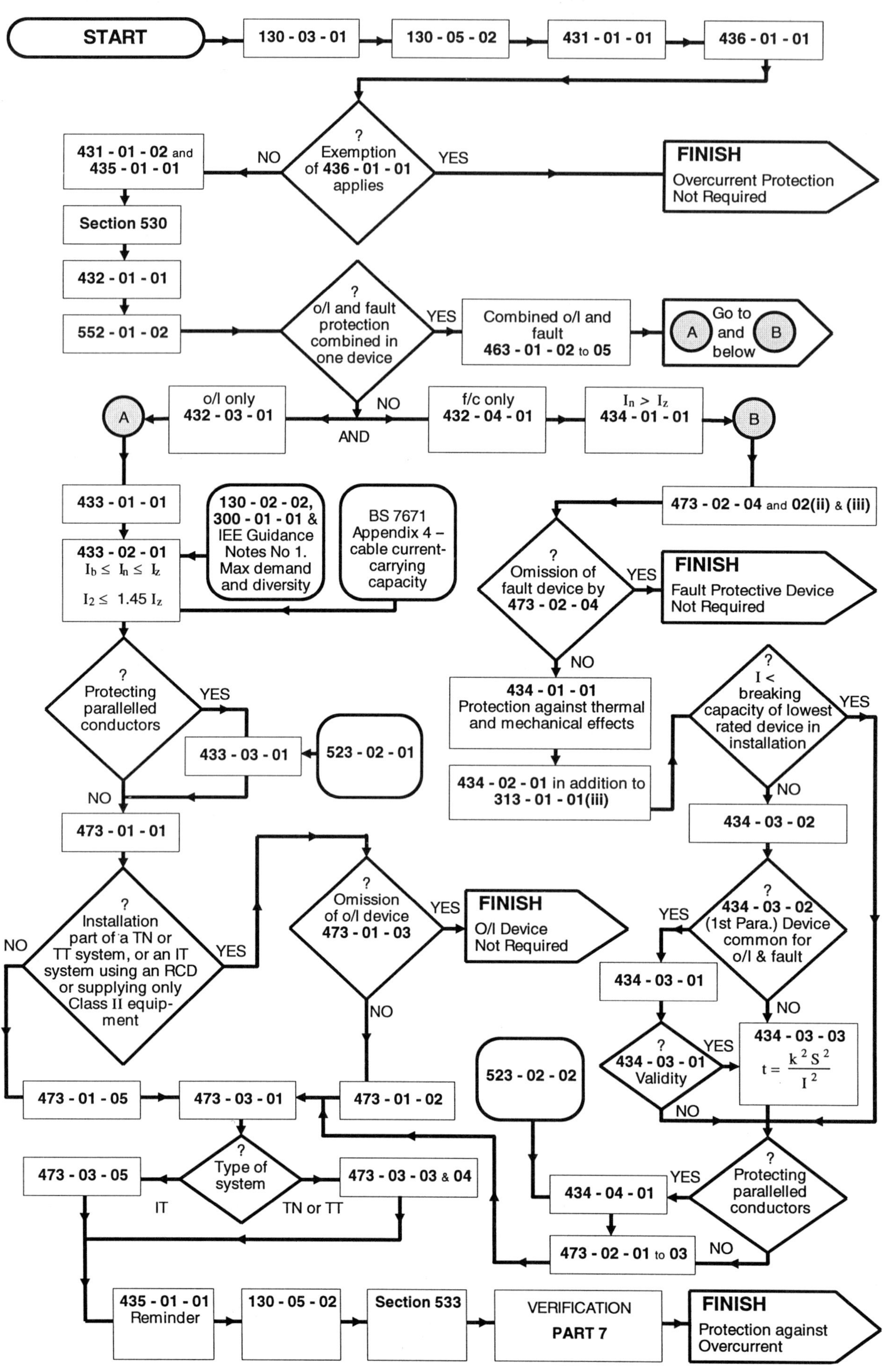

Protection against Overcurrent

Overcurrent is a current exceeding the rated value for a current-carrying part. It may be an overload current or a fault current.

Overload current occurs in an electrically sound circuit due to the connection of equipment requiring a current in excess of the design current. Short-circuit current flows when a fault of negligible impedance occurs between live conductors which are at different potentials under normal operating conditions.

Three other conditions are possible in addition to overload current and fault current as defined above:

(i) A fault of greater than negligible impedance, (e.g. tracking of insulation) occurring between live conductors, causes a condition similar to overload. Such a condition will either develop rapidly into a fault of negligible impedance and thus cause a short-circuit current to flow, or it will be cleared before this occurs by operation of the overcurrent protective device.

(ii) A fault occurring between a phase conductor and a protective conductor gives rise to an earth fault current of similar magnitude to a short-circuit current, and the overcurrent protective device will operate.

(iii) A fault, not of negligible impedance, between a phase conductor and a protective conductor may pass insufficient current to operate either an overload protective device or a fault protective device.

Conditions (ii) and (iii) are earth fault conditions and are covered in Topic Chart 2.

The overcurrent protective device provided for a particular circuit may not afford protection to the current-using appliance connected to that circuit, or to the flexible cables or cords which may supply it. This must be taken into account in the design of final circuits supplying non-fused socket-outlets. The designer is unlikely to be able to say with certainty what equipment will be connected at any one time. This is a matter to be assessed in accordance with Part 3 of BS 7671: 1992.

Protection against Overload Current

The rating of the overload protective device is related to the maximum temperature which the insulation of the conductors can withstand for a period of time without significant deterioration. Thus, overload protection is based on two fundamental requirements.

- the design current of the circuit must not be greater than the nominal rating or setting of the related overload protective device;
- the current-carrying capacity of circuit conductors must not be less than the nominal rating or setting of the related overload protective device.

These requirements may be expressed as:-

$$I_b \leq I_n \leq I_z$$

where I_b = design current of circuit;
I_n = nominal current or current setting of a protective device; and
I_z = current-carrying capacity of a cable.

Regulation 433 - 02 - 01(iii) requires the effective operating current of the device to be not greater than 1.45 times the current-carrying capacity of the related conductor, that is:

$I_2 \leq 1.45\ I_z$ where I_2 is the current ensuring effective operation of the protective device. For all protective devices except semi-enclosed fuses to BS 3036 (see page C6/3) this requirement is satisfied, if $I_n \leq I_z$.

An overload protective device is capable of carrying indefinitely its nominal current rating or setting, but above this value the speed of operation becomes faster as the current increases. In other words, the time of operation of the device is inversely proportional to the current flowing.

An overload protective device is capable of carrying a small sustained overload current for an appreciable period of time (one hour or more) depending on the rating and type of device, but advantage must not be taken of this when selecting the device. If small overloads are likely to occur frequently, they should be treated as part of the normal design current of the circuit. An example of this is where an electric motor is subjected to frequent starting. This is dealt with later under "Co-ordination of Overload and Fault Protection" (page C5/8).

Position of Overload Protective Devices

An overload protective device is normally required at each point along a conductor where its current-carrying capacity is reduced. Such reduction may be caused by a change in cross-sectional area, method of installation, type of cable or conductor, environmental conditions, etc. (see also Topic chart 6 and related notes).

The device protecting the reduced conductor is permitted to be placed along the run on the load side of the point of reduction of I_z, provided that no branches or outlets are introduced between the point of reduction and the device. For example, where a switch-fuse is mounted directly on to a busbar chamber, the conductors connecting the switch-fuse to the busbars may have a lower current rating than the busbars, but are regarded as protected against overload by the device which they are feeding, i.e. the switch-fuse. However, care must be taken that any reduced conductor is also protected against fault current (see below).

Particular attention is drawn to the IEE Guidance Notes No 1 Selection and Erection concerning the connection of a non-fused spur to a ring final circuit. A non-fused spur is permitted to feed only one single or one twin 13A socket-outlet, or one permanently connected load. Clearly, a 30A or 32A overcurrent protective device does not adequately protect a 2.5mm^2 cable in accordance with the requirements of Regulations 431 - 01 - 01, 433 - 01 - 01, 433 - 02 - 01, etc.

Acceptance of a non-fused spur having a cable the same size as that used for the ring, depends on the probability that such a spur will never be loaded beyond the current-carrying capacity of the cable. This assumes that the permitted twin 13A socket-outlet will never be subjected, for example, to two 13A loads simultaneously. This installation of non-fused spurs should be avoided, therefore, unless it can be determined that there is no likelihood of the non-protected spur cable being overloaded.

Regulation 473 - 01 - 04(ii) permits the installation of a consumer unit (c.u.) without overload protection of the tails connecting it to the Supplier's meter, provided that the aggregate maximum demand of the c.u. circuits, after any relevant diversity has been applied, is not likely to overload the tails.

Protection against Fault Current

The thermal effects of fault and short-circuit currents which can result in the melting of insulation and even fire are obvious, but the mechanical effects are often overlooked. Fault currents can produce magnetic fields which distort or break cable cleats and busbar supports. Hence, rapid disconnection from the supply is essential to prevent damage.

Determination of the prospective fault current may be made by measurement or calculation, or it may be ascertained from the Supplier who is required by the Electricity Supply Regulation 1988 as amended, to provide this information, but in no case is it likely to be a simple matter.

Measurement is applicable to energised supplies, and if used to compare with calculated or ascertained values, a useful data base will be acquired.

At the design stage, it will be necessary to calculate values based on the impedances of transformers and mains cables. However, the h.v. source impedance can be ignored and, where conductors of less than 35mm^2 csa are involved, the reactances can be disregarded and the calculation made on the resistance values only.

The value calculated is usually that of the prospective short-circuit current at the load side terminals of the overcurrent protective device at the origin of the installation, without any allowance for the impedance of the outgoing circuit or circuits.

The installation designer must use the best information available, and may even need to make his own estimate of the likely range of prospective fault current. Information is included in most manufacturers' literature and in Appendix F of BS 3036 for Semi-Enclosed Fuses.

The Electricity Association Engineering Recommendation P.25, gives details of short circuit characteristics of low voltage distribution networks, with particular reference to 230V single phase supplies up to 100A, with service cutouts having 100A fuses to BS 1361 Type II. Engineering Recommendation P.26 gives similar information for three phase supplies. Reference should be made to these publications (see part E of this Handbook) or to the relevant Supplier.

The following table gives selected values of prospective fault current for service cables and overhead lines up to 25mm^2 aluminium or 16mm^2 copper connected to supply networks other than those in the area of the London Supplier and in certain large cities, for which the prospective fault current should be assumed to be 16kA at the service cutout, irrespective of the length of the service line.

Length of Service Line (metres)	**Prospective Fault Current** kA
8	8.8
15	6.0
22	4.5
26	3.9
40	2.7

The values given relate to the prospective fault current at the cutout terminals and may be deemed also to apply at the consumers' main switchgear where this is located within approximately 1 metre of the cutout. Where the consumer's switchgear is remote from the cutout, for example in some multi-occupancy buildings, allowance may be made for the additional reduction of the prospective fault current due to the length of sub-main cable.

The length of the service cable may be assumed to be the distance from the cutout to the nearer kerb. (See Engineering Recommendations, P.26.)

Special attention is needed as regards tower blocks of flats where short lateral services may be supplied from substantial rising mains from a basement substation. The prospective fault current at the lower flats will be high but there is little likelihood of an additional substation being installed to increase the prospective fault current at the upper flats, although the original transformer could be increased in capacity to meet a growing demand.

Where the overcurrent protective device proposed for use at any point in the installation has a breaking capacity less than the prospective fault current at that point, it must be ensured that another protective device having adequate breaking capacity is installed on the supply side. The energy let-through of both devices must be co-ordinated so as to avoid damage to the load side device and to the conductors protected by the devices (see Regulation 434 - 03 - 02).

Although approximate, the prospective fault current for an installation may be determined by the use of a suitable instrument (provided the supply is available when the figure is needed), too much reliance should not be placed on such a reading. If it is low, there may be all the more reason for the supply to be reinforced at a later date, perhaps making switchgear inadequate for the new conditions. The prospective fault current should be the subject of discussion and agreement between the Supplier and the installation designer.

Although the Supplier's cutout fuse is not part of the installation, i.e. it is part of the source of energy, it may generally be used to provide overcurrent protection for the meter tails, and with a consumer unit, the busbar and other connections on the supply side of the overcurrent protective devices in the unit. Under these circumstances, the Supplier may specify a maximum length and minimum size of the meter tails, or may require additional protection against mechanical damage. The cutout may also provide the back-up protection necessary where the breaking capacity of the fuses or miniature circuit-breakers is less than the prospective fault current at their point of installation. BS 5486 Part 13 and BS EN 60439, Consumer Units, permits a manufacturer to allocate a PFC rating to the complete consumer unit when it is preceded by a 100A BS 1361 cutout fuse. For a consumer unit installed in a non-domestic situation, or not preceded by a BS 1361 fuse of 100A or less, then the advice of the manufacturer of the unit should be obtained. The meter tails must be in accordance with Regulation 473 - 02 - 02, i.e. not more than 3m long and enclosed in earthed steel or rigid pvc trunking – if the installation is protected by an rcd as part of a TT system, only rigid pvc trunking may be used.

The requirement that the prospective fault current shall be determined at every relevant point of the installation is automatically satisfied if the prospective fault current at the origin of the installation is less than the breaking capacity of the smallest rated device to be used in the installation.

A point to bear in mind is that the further away from the installation origin an overcurrent protective device is located, the greater will be the impedance of the related conductors, and the lower will be the prospective fault current at the protective device load side terminals. The breaking capacity of the fault protective device can be correspondingly reduced, therefore, but in practice, this may be unnecessary or undesirable, because of a wish to standardise the devices used in an installation and, consequently, their breaking capacities.

Fortunately, the wiring of most installations of a domestic or similar type will reduce the fault level to an acceptable value within a very short distance of the consumer's main switch, so that the probability of the consumer's protective device being subjected to a value of prospective fault current such that back-up protection is required to operate, is very low.

The Regulations permit the overcurrent protective device to be placed other than at the supply end of a cable and a particular fault protective device to be omitted provided that:

(i) an overcurrent protective device placed at the load end of the cable which affords overload current protection to the cable; and

(ii) the cable is protected against fault current by another device on its supply side.

This is shown diagrammatically as:

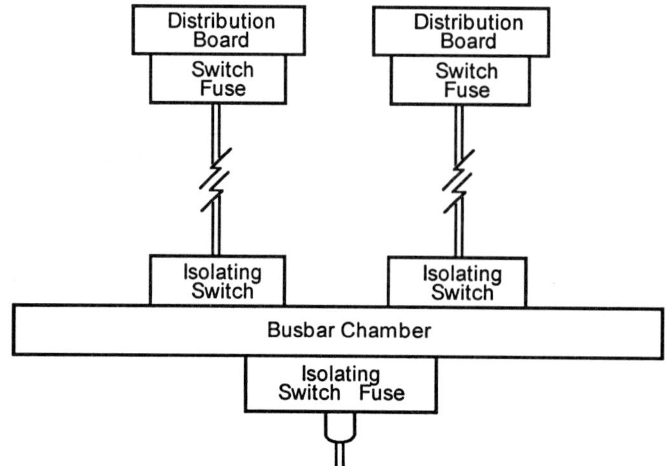

However, the required cable run may be of such impedance that there will be insufficient current at the load end to operate the supply side protective device. In addition, any subsequent variation in the supply network may result in a change in the prospective fault current, which may invalidate the fault protection.

Thus, it will generally be found that the arrangement shown diagrammatically below will be necessary, and the illustration in this section recognises this:

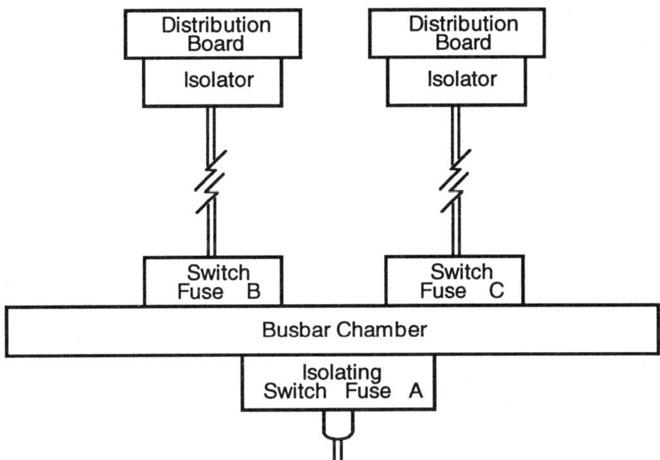

In the above example, devices of lower breaking capacity than the prospective fault current of the whole installation are permitted at points B and C provided that the device at A is suitable. However, it is essential to ensure that should a fault occur on one of the outgoing circuits, the energy let-through of A will not damage the particular device (B or C) or any conductors in the circuit. (Regulation 434 - 03 - 02.)

Overload current protection of the conductors between the busbars and the devices at A and C is dealt with above under Protection against Overload Current. A note on the meaning and application of energy let-through is given below.

Where larger installations and particularly industrial installations are concerned, prospective short-circuit current levels can be 10kA or more, and the rating of equipment used at the intake point must be able to deal with this. Information on the suitability of fault protective devices will usually be available from the manufacturer's literature on a particular device, or directly from the manufacturer when not contained in published information.

Some caution must be exercised if the use of imported equipment is contemplated. The test methods, and the presentation of results obtained, differ widely in various countries and often cannot be compared directly with British practice. Efforts are being made internationally to standardise test methods and parameters.

Where the overcurrent protective device complies with the requirement of Section 433 and has a rated breaking capacity not less than the prospective fault current at the point of installation, it may be assumed that the device is suitable for combined overload and fault protection, but it should be noted that not all circuit-breakers are suitable for this duty – although current-limiting types generally are. The contacts open rapidly, and the arc is dispersed so that both the rate of cooling and the arc's extinction are accelerated.

If a single device is to protect conductors connected in parallel, it is essential that the operating characteristics of the device be chosen carefully, particularly as regards energy let-through, bearing in mind that a fault may occur on one of the conductors. The I_z of such a circuit is the sum of the individual cable ratings (Regulation 433 - 03 - 01). Regulations 433 - 03 - 01 and 523 - 02 - 01 lay down conditions to attempt to ensure equal current sharing between conductors connected in parallel. However, it will still be difficult, if not impossible, to guarantee protection by the use of a single protective device, of each conductor of a number of conductors connected in parallel.

Co-ordination of Overload and Fault Protection

Attention has already been drawn to the need to co-ordinate the measures for overload and fault protection. This co-ordination is perhaps most familiar in the protection of motor circuits, where overload and fault protection are provided by separate devices, the fault device being located at the origin of the circuit and having an overrated nominal rating but suitable fault characteristic, whilst the overload protective device may be situated in the motor starter. Where, more commonly, a single device provides overload and fault protection, it must have characteristics suited to both functions.

Where an electric motor is subjected to frequent starting, the conductors and overload device must be related to the starting conditions of the motor, not to its running conditions. Infrequent starting of motors can be tolerated by overload protection devices, but careful attention must be given to suitable co-ordination of the characteristics of the associated fault protective device.

The Electricity Association Engineering Recommendation P28 states that frequent starting of motors is considered to be starting every two hours or less, but this is related to voltage dip only.

For most situations, advantage can be taken of Regulation 434 - 03 - 02 which allows the use of an overload protective device complying with Section 433 to protect the conductors on the load side of the device, provided that it has a rated breaking capacity not less than the prospective fault current at the point of installation.

Circuit Limitations

To ensure the protection of conductor insulation against thermal damage during fault conditions (i.e. the energy withstand of the cable, k^2S^2, must not be less than the energy let-through of the device, I^2t) the requirements are:

(i) for a fuse - the PFC must be sufficiently high (i.e. live conductors must have a correspondingly low impedance) to ensure a short enough disconnection time and

(ii) for a miniature circuit breaker - the same condition applies but, because of the shape of the characteristic curves of these devices, there is also a maximum fault current which must not be exceeded (i.e. live conductors must provide adequate impedance to limit the current to below the breaking capacity of the device). It follows that great care must be excercised if installation and/or modifications require alteration to lengths of conductors protected by miniature circuit-breakers.

Discrimination and Energy Let-through

Discrimination between all protective devices, whether for overcurrent or shock protection, must be ensured in order to meet the requirements of Regulation 314 - 01 - 02 and 533 - 01 - 06, i.e. to avoid danger and minimise inconvenience in the event of a fault. Discrimination must take into account any fault currents and overload currents which may occur.

The main factor which governs the stress and damage to a conductor is the specific energy let-through of the protective device. The values for a particular device are usually specified in ampere-squared seconds (A^2s) and referred to as I^2t let-through.

Manufacturers provide values of I^2t for their devices, and the value for a particular device must not raise the cable conductors to their limiting temperature before disconnection occurs. The method of calculation of the time taken to raise the conductors from the highest permissible temperature in normal duty to limit temperature is given in Regulation 434 - 03 - 03.

The values of energy let-through are also used in assessing discrimination between overcurrent devices in series. In general, if the total energy let-through of the load side device is less than the pre-arcing energy let-through of the supply side device, then discrimination will be obtained. For example, if a fuse-link has a minimum pre-arcing energy let-through of $20kA^2s$, to obtain discrimination on fault, a down-stream device should have a total energy let-through less than $20kA^2s$.

Manufacturers' advice should be obtained if consideration of I^2t values suggest a less than 2:1 ratio between nominal ratings of series fuses. However, the breaking capacity of any device must be related to the prospective fault current at its point of installation except where advantage is taken of Regulation 434 - 03 - 01, as in the example above (page C5/6).

Careful assessment of characteristics is necessary to ensure discrimination between fuses and miniature and moulded case circuit-breakers and between circuit-breakers from different manufacturers.

(a) Fuses

Discrimination is necessary to ensure that a fault will be cleared by the appropriate circuit fuse without affecting other fuses through which it is supplied (Reg 533 - 01 - 06). If fuses have not been selected to achieve discrimination, then although faults will be cleared, it will only be at the expense and inconvenience, at the very least, of replacing more than one fuse or sets of fuses. There will be proper discrimination under the worst conditions, i.e. with the highest attainable fault current, when the total I^2t of the smaller (minor) fuse is less than the pre-arcing I^2t of the larger (major) fuse. Under high fault current conditions fuse operation occurs in less that 0.1 second but with lesser faults taking more than 0.1 second to operate the fuse, reference must be made to the manufacturer's published data for the pre-arcing times of the fuses. The foregoing refers in particular to HBC fuses to BS 88 or 1361, but is still broadly true for semi-enclosed (rewirable) fuses to BS 3036, should their somewhat restricted fault capacity be acceptable for a particular situation. It should be noted that discrimination may fail where, prior to the occurrence of the fault, the major fuse has been heavily loaded, or has deteriorated due to repeated transient overloads.

(b) Circuit-Breakers

It may be more difficult to ensure discrimination with either circuit breakers or moulded case circuit breakers. This is because at lower (overload) currents, circuit-breakers operate with an inverse time delay due to their thermal mechanisms, but with higher (fault) currents, the magnetic action over-rides the thermal action resulting in almost instantaneous tripping (less than 0.1 second).

MINATURE CIRCUIT BREAKER TO BS 3871 AND CIRCUIT BREAKERS TO BS EN 60898 – MAIN FEATURES					
Type	**Nominal rating, I_n (A)**	**Overload characteristic***	**Current band causing instantaneous operation, I_{inst}**	**Current necessary for instantaneous operation**	**Common utilisation**
1	$10 \geq I_n$	$1.50I_n$	$2.7I_n$ to $4.0I_n$	$4.0I_n$	General circuits where load does not exhibit high inrush characteristics, eg. tungsten lighting loads.
	$10 < I_n$	$1.35I_n$			
B	All	$1.45I_n$	$3.0I_n$ to $5.0I_n$		
2	$10 \geq I_n$	$1.50I_n$	$4.0I_n$ to $7.0I_n$	$7.0I_n$	General circuits where load only exhibits moderate inrush characteristics, eg. large tungsten and fluorescent lighting loads.
	$10 < I_n$	$1.35I_n$			
C	All	$1.45I_n$	$5.0I_n$ to $10I_n$	$10I_n$	General circuits where load exhibits moderate to high inrush characteristics, eg. motor loads, air conditioning plant etc.
3	$10 \geq I_n$	$1.35I_n$	$7I_n$ to $10I_n$	$10I_n$	
	$10 < I_n$	$1.35I_n$			
D	All	$1.45I_n$	$10I_n$ to $20I_n$	$20I_n$	General circuits where load exhibits high to harsh inrush characteristics, eg. X-ray equipment, welding equipment, DOL motors, circuits with transformers, etc.
4	$10 \geq I_n$	$1.50I_n$	$10I_n$ to $50I_n$	$50I_n$	
	$10 < I_n$	$1.35I_n$			

*** Current required for overload protection within 'conventional time' at reference temperature. Conventional time for Types 1, 2, 3 and 4 and for Types B, C and D up to $I_n \leq$ 63A is 1 hour. For Types B, C and D of $I_n \geq$ 63A the conventional time is 2 hours.**

Discrimination can therefore only be attained when the instantaneous tripping current of the smaller circuit- breaker does not approach that of the larger. Account has also to be taken of manufacturing variations in items of the same manufacture, and the different characteristics of other makes and designs. Where circuit-breakers are to be connected in series therefore, the manufacturers, or their data, should be consulted.

(c) Circuit-Breakers with Back-up Fuses

When circuit-breakers are supplied through fuses, sizing should be such that for fault levels up to about 1kA, reliance is placed on the circuit-breaker to clear the fault – and this value will cover the majority of final circuit faults. For higher current levels the fuse will take over, although the change-over point will depend on the current ratio of the two components, and also on the fuse characteristics. However, the use of energy limiting circuit-breakers can reduce the need for fuse back-up, but their shorter operating times may make discrimination difficult, and again, the manufacturer's advice should be sought.

Topic Chart 6

A DECISION CHART

Live Conductors

Current-Carrying Capacity of a Conductor is the maximum current which can be carried continuously by a conductor under specified conditions, without its steady state temperature exceeding a specified value.

Design Current (of a Circuit) is the magnitude of the current intended to be carried by the circuit in normal service (rms value for a.c.).

Live Part means a conductor or conductive part intended to be energised in normal use, including a neutral conductor, but by convention, not a PEN conductor.

Hazardous-live-part. A live part which can give, under certain conditions of external influence, an electric shock.

In the Topic Chart, conditions to be satisfied include the following quantities:

I_b = design current of circuit
I_n = nominal current or current setting of a protective device (fuse or overcurrent circuit-breaker)
I_z = current-carrying capacity of a cable (when installed)
I_t = tabulated single-circuit current-carrying capacity of a cable
C_g = correction factor for grouping
C_a = correction factor for ambient temperature
C_i = correction factor for conductors embedded in thermal insulation
C_t = correction factor for operating temperature of conductor (^{o}C)
C_d = correction factor for semi-enclosed (BS 3036) fuses is 0.725, for other devices, 1.0

See Part 2 of BS 7671: 1992 for the complete set of Definitions

TOPIC CHART 6 (Decision) Live Conductors

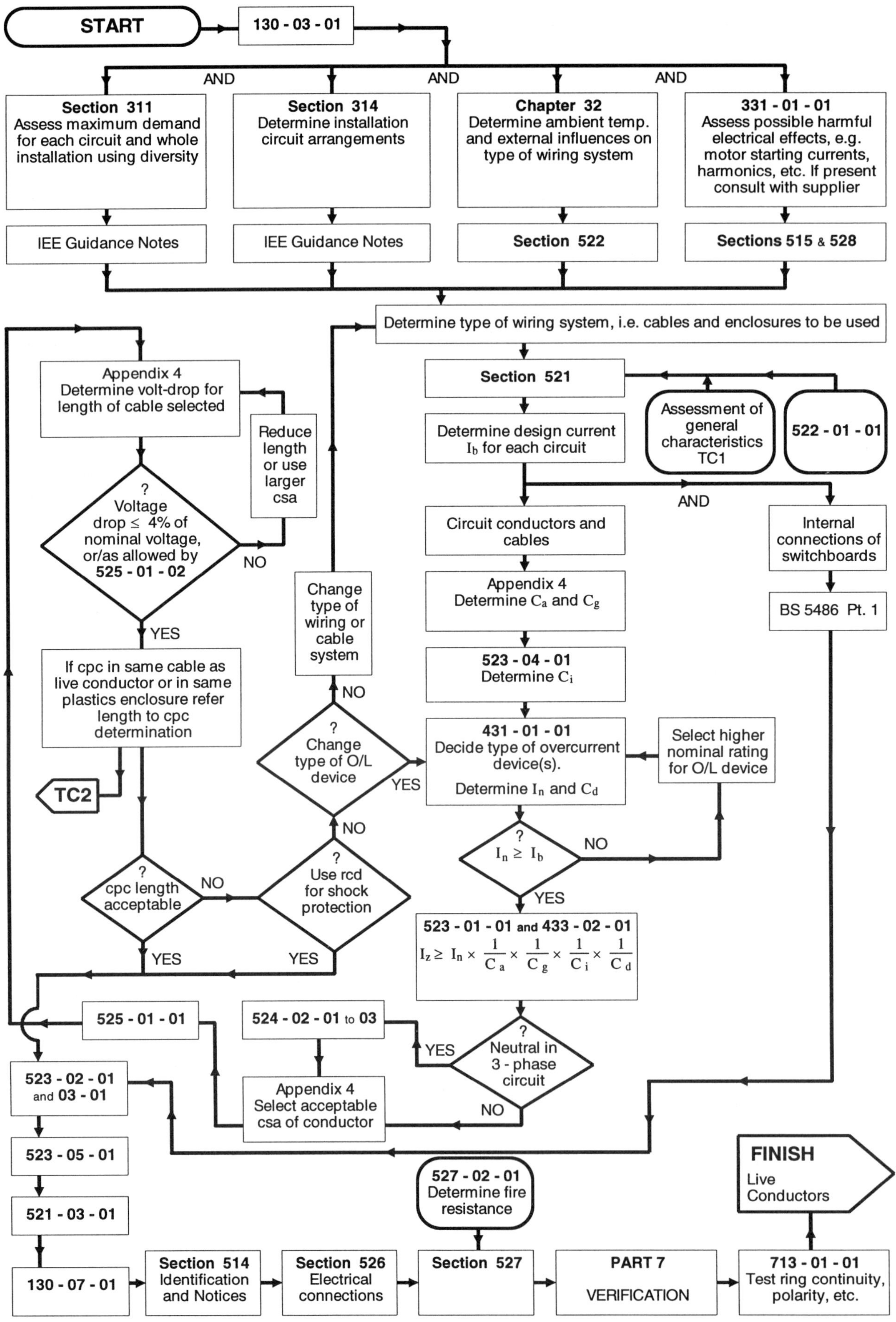

Live Conductors

The minimum permissible cross-sectional area of the live conductors (phase and neutral) of any circuit depends on:-

– design current of the circuit (I_b)
– type of overcurrent protective device and its nominal rating or setting (I_n)
– type of cable
– method of installation and environmental conditions
– thermal constraints under fault conditions
– earth fault loop impedance limitations

This chapter deals only with the selection of cables on the basis of current-carrying capacity.

As stated on page C5/3 of this Handbook, the relationship between I_b, I_n and I_z may be expressed as

$$I_b \leq I_n \leq I_z$$

Where an overload protective device has been omitted in accordance with Regulation 473 - 01 - 03 and 04, no value can be assigned for I_n so that the relationship between I_b and I_z becomes

$$I_b \leq I_z$$

For any circuit, it is necessary, therefore, to select a cable having a current-carrying capacity (I_z) when installed, which is equal to or greater than I_n or I_b as appropriate.

The values of current-carrying capacity, I_t, given in Tables 4D1A to 4L4B of BS 7671: 1992 apply only when a cable is installed under the conditions stated in the headings to the relevant columns of each Table. In those circumstances $I_z = I_t$, and it is, therefore, a simple matter to select a cable having the required current-carrying capacity.

Where the conditions of installation differ from those on which the Tables are based, the effective current-carrying capacity of the cable may vary, thus necessitating the selection of a cable of different cross-sectional area.

The following procedure is recommended for the selection of cables when overload protection is required:

1. Determine the design current of the circuit I_b
2. Select the type and rating of the overload protective device $I_n \leq I_b$
3. Select the type of cable and refer to the appropriate Table 4D1A to 4L4B
4. Determine the rating factor for groups, using Table 4B1 C_g
5. Determine the rating factor for ambient temperature using either the correction factors in Table 4C1 where semi-enclosed fuses are not used or Table 4C2 where semi-enclosed fuses are used C_a
6. Determine the rating factor for thermal insulation (see below) C_i
7. If semi-enclosed fuses have been selected and if the cables are not mineral insulated, use factor 0.725. For other types of overcurrent protective device and for mineral insulated cables, the factor is 1.0 C_d

8. Determine the tabulated current-carrying capacity -

$$I_t \geq I_n \times \frac{1}{C_g} \times \frac{1}{C_a} \times \frac{1}{C_i} \times \frac{1}{C_d}$$

9. From the appropriate Table 4D1A to 4L2B, select a cable having the value I_t determined at step 8

It should be noted that the effective current-carrying capacity, I_z of the selected cable, when installed, is equal to $I_t \times C_g \times C_a \times C_i \times C_d$

When overload protection is not required, step 2 is omitted and step 8 becomes:

8. Determine the tabulated current-carrying capacity –

$$I_t \geq I_b \times \frac{1}{C_g} \times \frac{1}{C_a} \times \frac{1}{C_i}$$

Thermal Insulation

Wherever possible, cables should be routed clear of any space where thermal insulation may be installed. Where this is not possible the current-carrying capacity must be reduced.

Appendix 4 of BS 7671: 1992 gives correction factors for cables installed in accordance with Method 4, Appendix 4.

Where single cables are directly enclosed in thermal insulation for more than 0.5m of their length, their current-carrying capacity shall be taken as 0.5 times the tabulated current-carrying capacity for that cable installed in accordance with Method 1, Appendix 4.

Where the cable is only enclosed in thermal insulation for a short distance, i.e. for less than 0.5m, the correction factors of Table 52A apply.

Where a cable is buried in thermal insulation but is in contact with a thermally conducting surface (e.g. structural steelwork), the rating factor C_i may be taken from Appendix 4.

Where a cable is installed in accordance with Method 1 of Table 4A, resting on rigid thermal insulation, no further derating is necessary.

Multiple Application of Rating Factors

When determining current-carrying capacities, it is important to remember that unless the circumstances for reduction in capacity apply simultaneously to a cable, then having derated for the worst situation, no further reduction is necessary. For example, a group of eight circuits of single-core cables running for some distance out from the distribution board would be subject to a grouping factor of 0.5 (Table 4B1). If the cables of these circuits continue either grouped or as a single circuit for more than 0.5m surrounded by thermal insulation, the appropriate factor, 0.5, (Reg. 523 - 04 - 01) would have had to be applied and the resulting multiplier would be 0.25 (i.e. 0.5 x 0.5). However, if only single cables are surrounded by thermal insulation, then the 0.5 grouping factor applied at the start of the circuits already provides the 0.5 derating required by the thermal insulation.

Thermal Protection of Phase Conductors under Earth Fault Conditions

A situation can arise where an installation apparently in accordance with the Regulations may be lacking in thermal protection of phase conductors during earth fault conditions. This can happen if a circuit is protected against fault, but not overload, at its source, e.g. a distribution board, with overload protection some distance away, e.g. combined with motor control gear. This is a permissible arrangement under 473 - 01 - 01 and 02.

Where fault and overload protection are combined then 434 - 03 - 02 applies. For the situation outlined above, 434 - 03 - 03 is applicable and the adiabatic equation must be satisfied. If the maximum fault current is above that corresponding to the intersection of the adiabatic curve for the phase conductor(s), and the time/current characteristic of the overcurrent protective device, then 434 - 03 - 03 is satisfied, but if the overcurrent is less than this value, the clearance time is greater and the conductors are not thermally protected.

Such a situation can arise when the earth fault current is less than the short-circuit current, caused e.g. by reduced cross-section of protective conductors provided under Table 54G.

With TN-S and TT systems there is every likelihood that earth fault currents will be less than short-circuit currents. In all the circumstances described, the phase conductors as well as the protective conductors must be checked for adequacy against thermal damage.

Attention is therefore drawn to Regulation 543 - 01 - 01 and to Regulation 434 - 03 - 03.

However, where the impedance of the earth return is less than that of the neutral conductor, as may occur with mineral insulated copper covered cables, then the earth fault current can exceed the short-circuit current.

Figure C10

Chapter 52
Selection of Types of Wiring Systems

N.B. This illustration does not attempt to indicate all of the specific permissible types of cables, conductors and material for particular situations referred to in Chapter 52

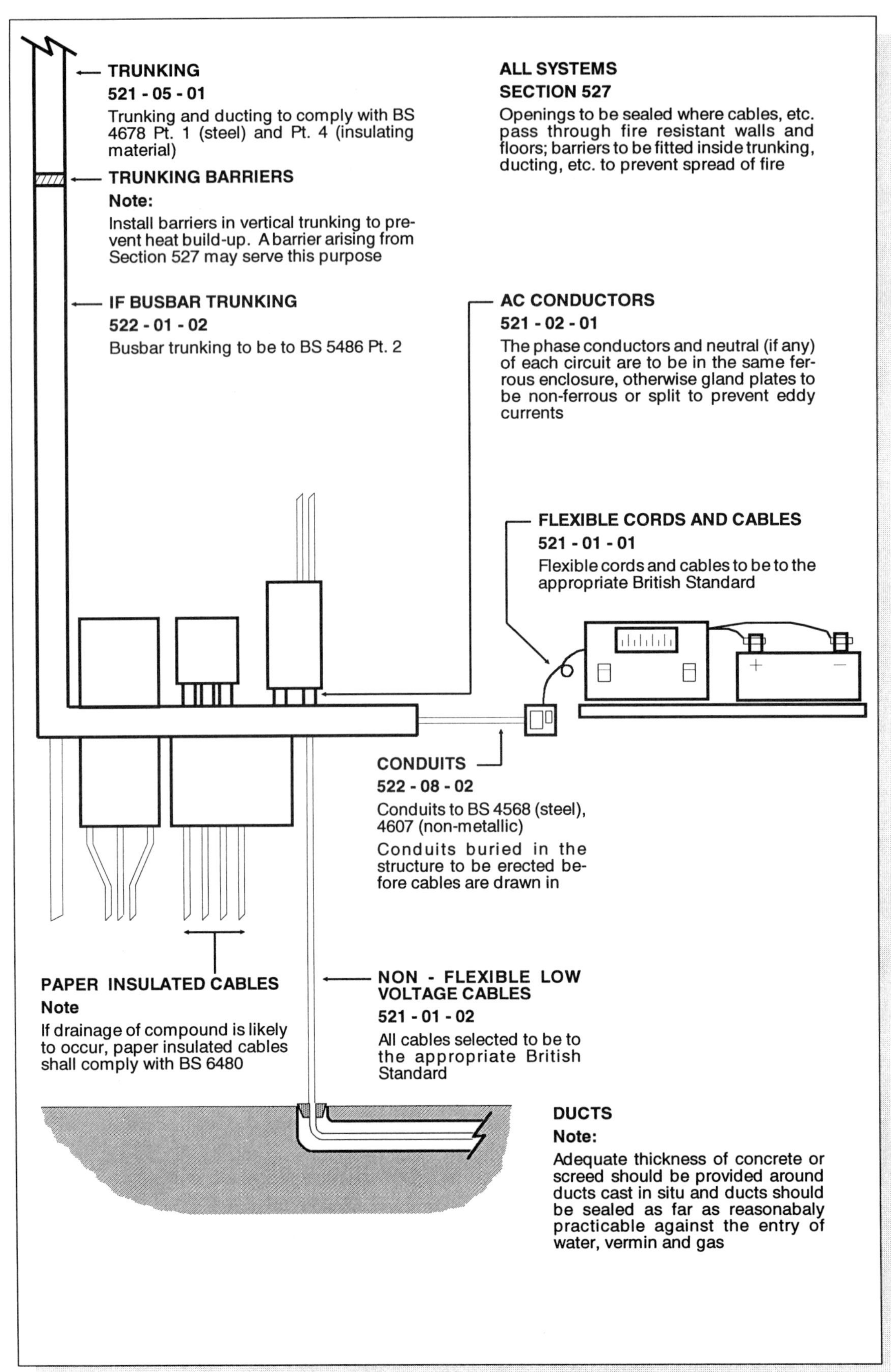

Figure C11

Chapter 52
Operational Conditions:
Current-Carrying Capacity

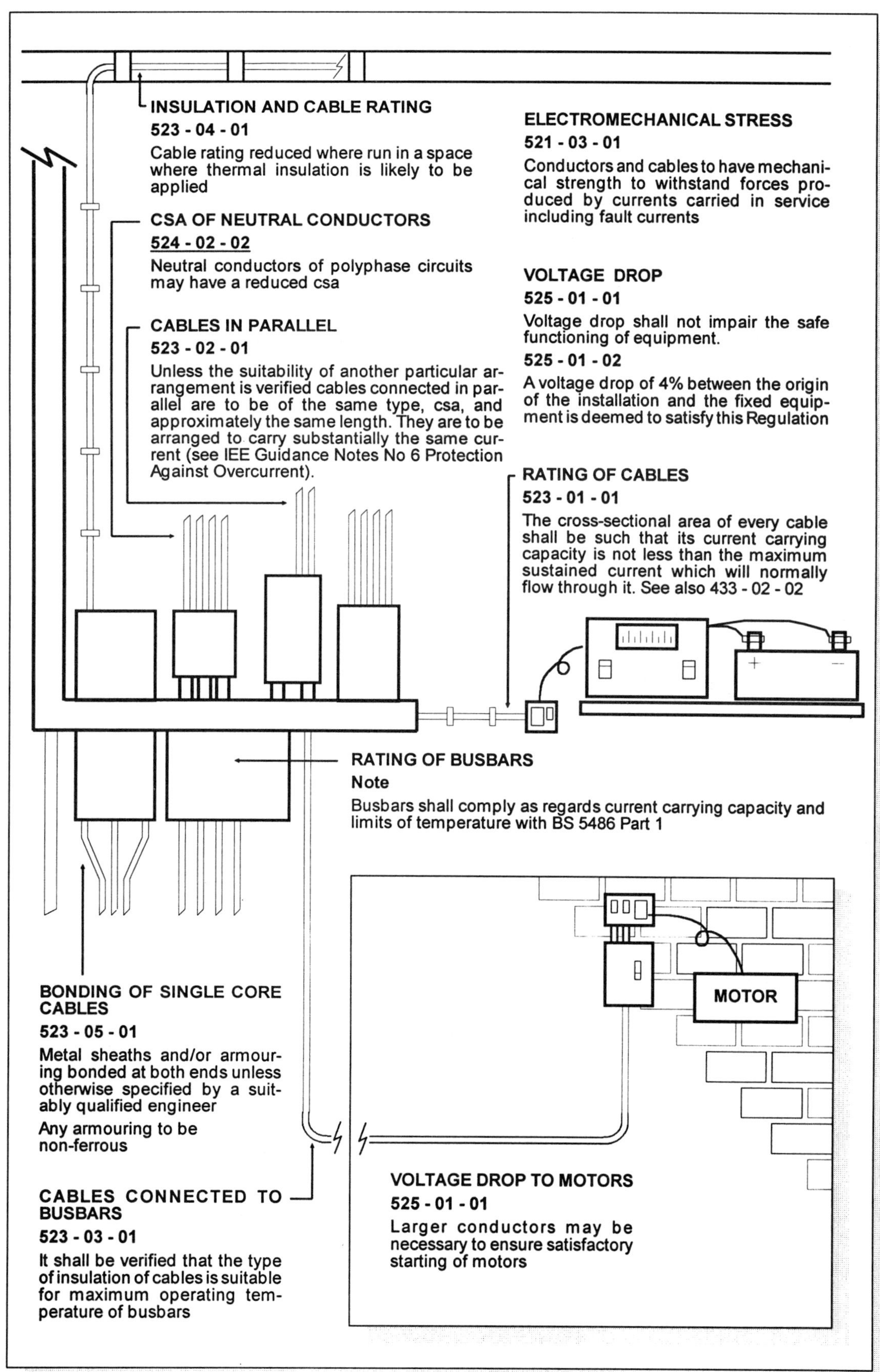

Protection against Thermal Effects and External Influences

Ambient Temperature is the temperature of the air or other medium where the equipment is to be used.

External Influence is any influence external to an installation which affects the design and safe operation of that installation.

See Part 2 of BS 7671: 1992 for the complete set of Definitions

TOPIC CHART 7 (Information) Protection against Thermal Effects and External Influences

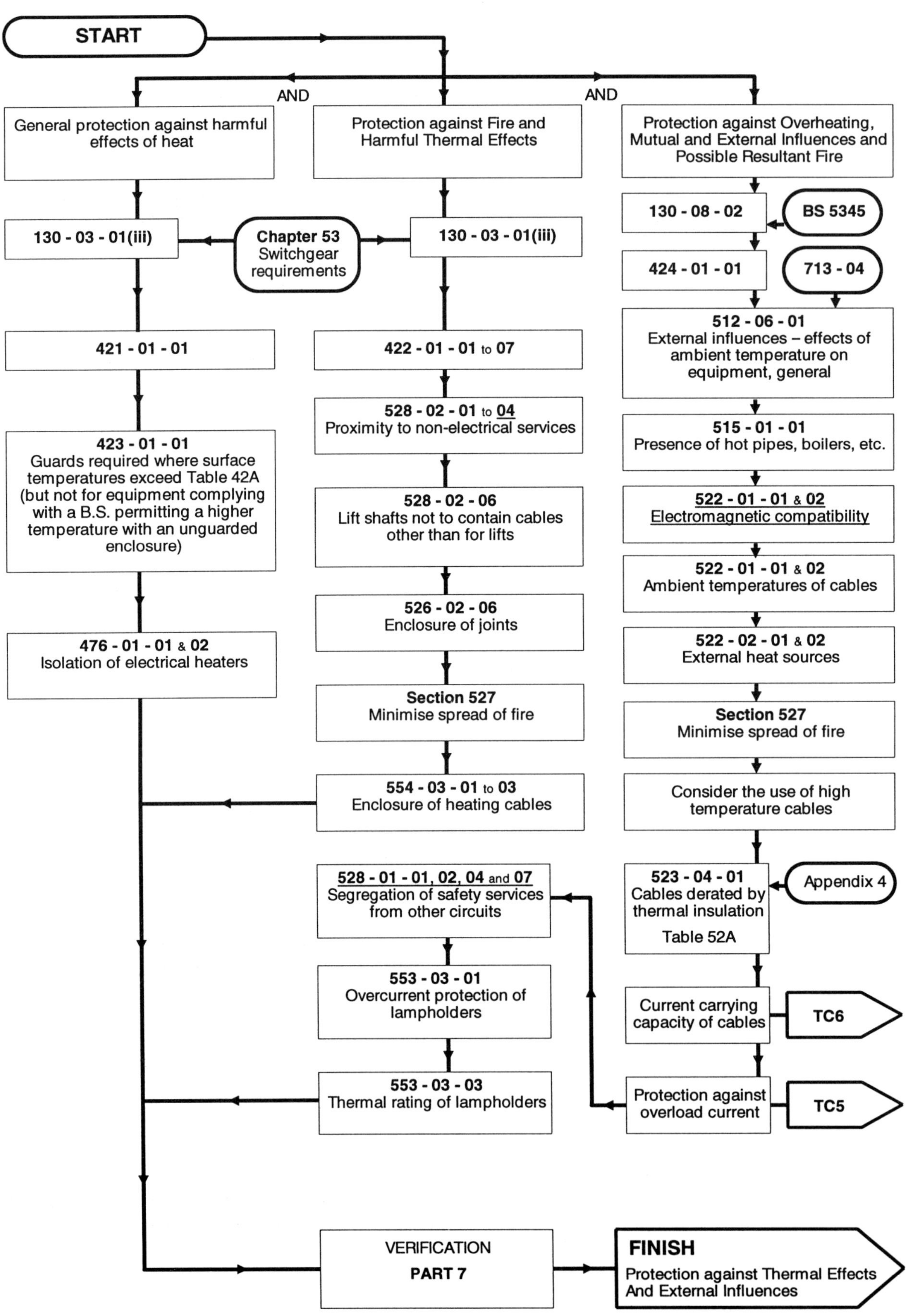

Protection against Thermal Effects and External Influences

Topic Chart 7 draws together all the Regulations concerned with thermal effects, whether produced by the electrical equipment itself in normal service or due to external causes. In addition, some of the precautions to be taken also assist in restraining the spread of fire through the premises concerned, and notes on these are also included.

Ambient Temperature is the temperature of the air or other medium in which equipment is to be used. The ambient temperature of an item of equipment is the temperature at the position where the equipment is to be installed, resulting from the influence of all other equipment in the same location, when operating, but not taking into account the heat generated by the item of equipment itself.

Generally, an item of equipment manufactured in compliance with a British Standard will, where relevant, be specified for operation at a maximum ambient temperature, the design of the equipment being such that the heat generated within it is safely dissipated.

External Influence is any influence external to an installation which affects the design and safe operation of that installation. It may affect a particular item of equipment, a circuit, or a whole installation, and may be an influence within or outside a building. For examples of external influences, refer to Section 522 and in particular to codes AA, BE2, CA and CB2, relating to thermal influences.

Protection against Thermal Effects Produced by Electrical Equipment

The thermal effects dealt with here are those arising from the equipment in normal service. Additional thermal effects caused by overcurrent (Topic Chart 5), overload current and fault current are not considered.

Protection against Fire

The Regulations identified in the chart under this heading are those intended to minimise the risk of fire caused by the electrical installation.

Regulations 528 - 02 - 01 to 06 define the requirements for the installation of a wiring system in "Proximity to Non-Electrical Services". Compliance will reduce the likelihood of a fire due to a leak from other services in conjunction with the operation of electrical services.

Cables other than those associated with lifts must not run within a lift shaft, not only because of the risk of heat building up in the shaft but also because of the risk of mechanical damage to the cables, and the difficulties associated with inspecting and maintaining them. (528 - 02 - 06).

Regulation 526 - 01 - 01 deals with the soundness of all cable terminations and joints, the obvious reference being to the possibility that a loose or unsound joint could cause an excessive temperature rise, arcing or fire.

Similarly, Regulations 526 - 03 - 02 and 03 refer to the enclosure of terminations or joints in suitable protective material. The requirement for material complying with characteristic "P", BS 476, Part 5. Characteristic "P" indicates that the material has passed the specified test and attention is drawn to the fire terminology.

The fire sealing required under Regulation 527 - 02 - 01 is specifically required to restrict the spread of fire which might otherwise occur through conduits, ducts, etc. An obvious example is a rising main, whether it be comprised of cables or busbars enclosed in metal or cables fixed directly to walls.

A reference is made in Fig. C10 page C6/6 to thermal barriers which may be required to prevent unacceptable temperatures existing at the top of channel duct, ducting or trunking. These can be the fire barriers, if fitted. Thermal barriers may not have the required degree of fire resistance to act as fire barriers,and if barriers are provided in proprietary rising main equipment, their fire or thermal rating must be ascertained.

Topic Chart 8

AN INFORMATION CHART

Selection and Erection of Equipment

TOPIC CHART 8 (Information) Selection and Erection of Equipment

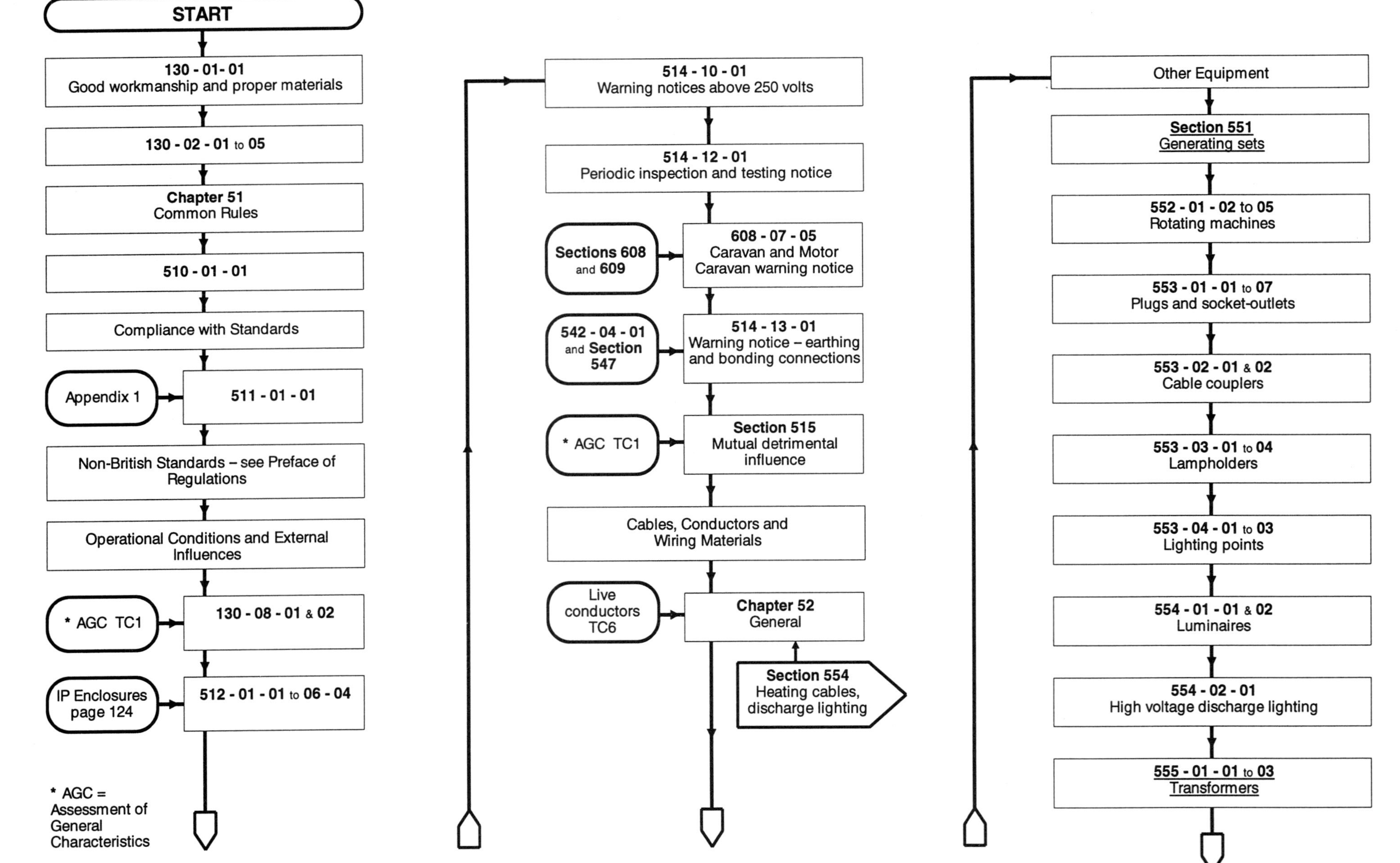

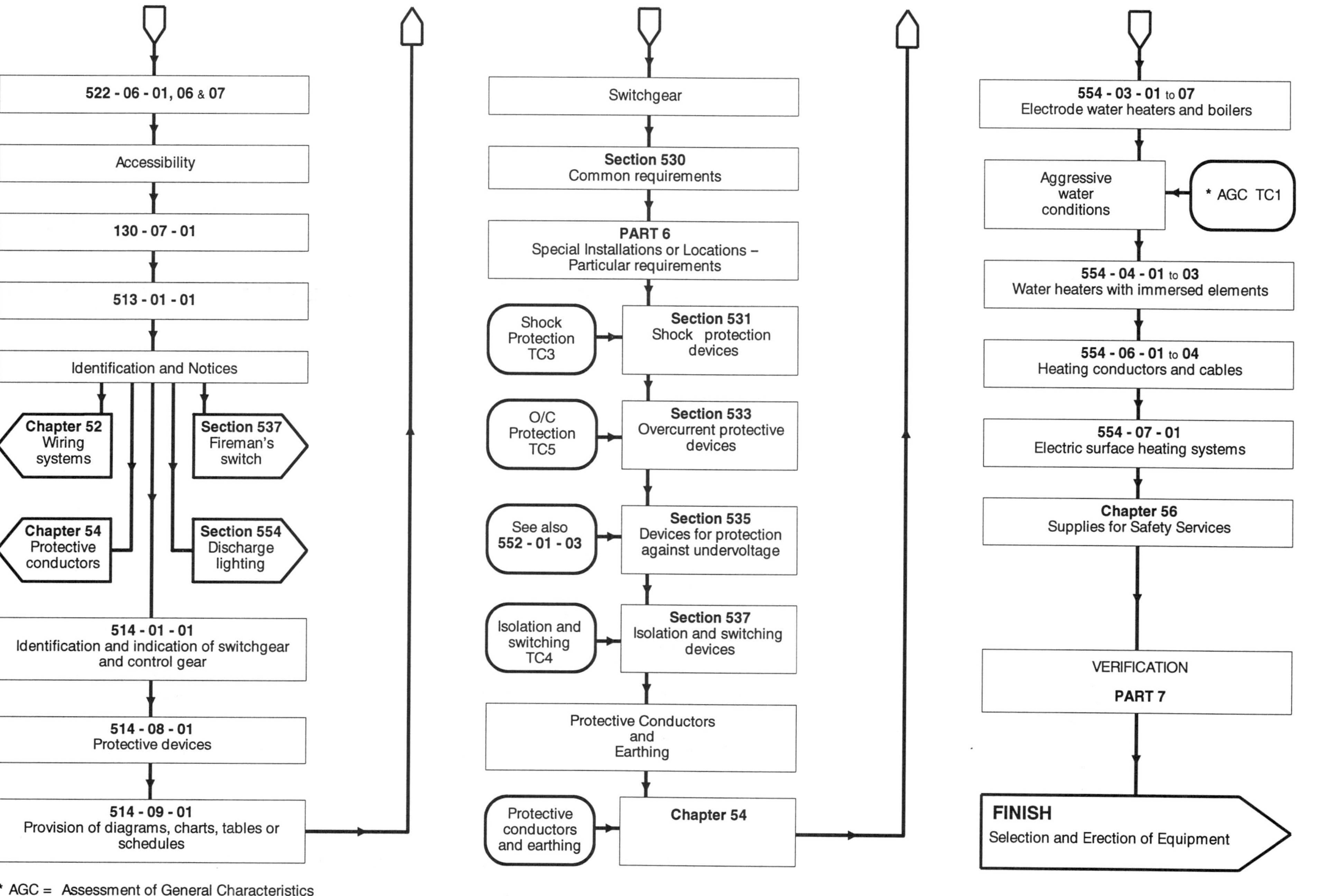

522 - 06 - 01, 06 & 07
Accessibility
130 - 07 - 01
513 - 01 - 01
Identification and Notices
Chapter 52 Wiring systems
Section 537 Fireman's switch
Chapter 54 Protective conductors
Section 554 Discharge lighting
514 - 01 - 01 Identification and indication of switchgear and control gear
514 - 08 - 01 Protective devices
514 - 09 - 01 Provision of diagrams, charts, tables or schedules
* AGC = Assessment of General Characteristics
Switchgear
Section 530 Common requirements
PART 6 Special Installations or Locations – Particular requirements
Shock Protection TC3
Section 531 Shock protection devices
O/C Protection TC5
Section 533 Overcurrent protective devices
See also 552 - 01 - 03
Section 535 Devices for protection against undervoltage
Isolation and switching TC4
Section 537 Isolation and switching devices
Protective Conductors and Earthing
Protective conductors and earthing
Chapter 54
554 - 03 - 01 to 07 Electrode water heaters and boilers
Aggressive water conditions
* AGC TC1
554 - 04 - 01 to 03 Water heaters with immersed elements
554 - 06 - 01 to 04 Heating conductors and cables
554 - 07 - 01 Electric surface heating systems
Chapter 56 Supplies for Safety Services
VERIFICATION PART 7
FINISH Selection and Erection of Equipment

Selection and Erection of Equipment

Good workmanship is a requirement of the Regulations, not merely as regards the practical aspects of installation work, but also concerning the fitness of the work being done for its ultimate purpose.

It is anticipated that every installer and designer will need at least access to, if not possession of, the various British Standards referred to by name in the Regulations. Sometimes the information gained from reading the British Standard may seem academic, but the responsibility for the correct selection and use of equipment rests with the designer and installer. Consequently, it is important to note at least the way the various British Standards have been taken into account.

For example, consider BS EN 60947, which deals with low voltage motor starters. Regulation 435 - 01 - 01 draws attention to the fact that the particular short circuit co-ordination and arrangements of the Standard are not precluded by the Regulation. BS EN 60947 grades the acceptable damage to starters under faul conditions with an acceptable risk of contact welding. Clearly, the designer or installer ought, at least, to be aware of the possible condition in which equipment may be, following a fault. The performance of the starter should be borne in mind when assessing the importance of the running of the system.

Another example is BS 5266 which incorporates compliance with the Regulations and in particular the volt drop, inspection and testing requirements of emergency lighting installations, all of which are to be taken into consideration by designers and installers.

It is important to bear in mind that any manufacturer may claim compliance with British Standards and, should such a claim be questioned in a particular case, it is necessary to produce evidence that the claim was made by the manufacturer. Thus, it is better (when possible) to use equipment which is marked as to its compliance.

There are certain duties of the Health and Safety at Work Act, Section 6 which are effected by the Consumer Protection Act which concern designers and installers. These are to ensure that all necessary research is done in design or manufacture to eliminate or minimise risk, and to ensure that there is nothing unsafe about the way in which the "article" is erected or installed.

However, the Act does not require the designer/manufacturer/installer to repeat tests or research already done by others, if it is reasonable to rely on the previous tests. As an example, equipment bearing the BSI Safety Mark would probably come under this heading, as far as the inherent safety of that piece of equipment is concerned.

When making a selection of equipment or deciding upon the method of erection protection against damage must be incorporated. This may be achieved by the correct selection of enclosure or by location.

Regulation 522 - 06 - 01 requires that all conductors and cables be protected against mechanical damage to which they may be subject in normal service. This protection can be separate from the conductors, e.g. conduit, or combined with them, e.g. wire armouring. Where cables are concealed within walls or partitions, Regulation 522 - 06 - 06 requires that cables be run within 150mm of the tops of the walls or partitions, or within the same distance of vertical corners. To reach outlets and accessories outside these zones, cables must be run straight, and either vertically or horizontally. Unearthed metallic or plastic capping may be used to protect the cables during plastering or other site operations.

Regulation 522 - 06 - 07 provides that when compliance with 522 - 06 - 06 is impracticable, then either an approved cable (i.e. one having an earthed metallic covering or be of insulated concentric construction) is to be used, or the cable is to be enclosed in earthed conduit, trunking or ducting, which must itself satisfy the requirements for a protective conductor in that situation.

Enclosures are selected according to the protection they give against contact with live conductors either by a part of the body or a conducting material and by the protection they give against the ingress of dust or water. They are known as IP ratings and are detailed on page C8/7.

In addition the selection must take full account of Section 522 which states the range of external influences to be considered.

The degree of maintainability to be afforded to equipment could lead to dispute, in that the position of equipment might well present difficulties for the maintenance electrician.

Obviously, it is necessary to consider what is reasonable. It may be, for example, that luminaires over a swimming pool require catwalks alongside to facilitate maintenance (usually these would be within a false ceiling space with top access luminaires). Clearly, to have to erect scaffolding spanning the pool would be unreasonable when wishing to clean or maintain the fittings. On the other hand, it may be perfectly reasonable to assume that special access equipment is available for external discharge lighting or advertising signs, but the intention to provide it must be verified.

With the correct choice of equipment sited in an acceptable situation with regard to maintenance, there remains the task of adjusting and co-ordinating the equipment (where such equipment has this facility) to the required degree of protection and the actual installation conditions.

Clearly, the task of final adjustment must carry with it the work of recording the information and characteristics of equipment so adjusted, as well as providing suitable warning notices with respect to the use and purpose of the equipment. This is specifically covered by Regulation 514 - 09 - 01 and in Part 6.

Some information contained within the section of the Regulations concerned with labelling and provision of warning notices is vital to the subject described in the notice, as the requirement is not mentioned elsewhere in the Regulations.

All those concerned with work on building sites will be aware of the problems relating to the final positioning of services, causing routes of pipes, cables, etc., to be modified and redesigned as the building work progresses. With this in mind, it is essential that adequate supervision be given to the installation of cables and conductors, together with frequent verification of the performance of the earth path in terms of impedance values. During the inspection and testing required by Part 7, the adequacy of equipment to disconnect within either 0.4 or 5 seconds must be determined.

A new regulation 522 - 12 - 03 requires any penetration of load bearing structures by a wiring system to preserve the load bearing capacity. This means liaison with structural engineers is needed to obtain approval before any penetration of a load bearing structure is carried out.

Certain specific requirements are made in Chapter 55 under the heading of 'Other Equipment'.

Although not a requirement, inductors and transformers should be placed as near as practicable to the associated lamp. This is important for maintenance, and it may also assist in the efficient striking and functioning of the lamp. Mineral insulated cable should not be used to interconnect such lamps and their controlgear, since it can be damaged by the high striking voltages which may be encountered.

Plugs and sockets to BS 1363 and BS 546 shall be used only in locations where the peak ambient temperature will not exceed 35°C, and the average temperature will not exceed 25°C over 24 hours.

Whilst most of the considerations of selection and erection of equipment is related to it being connected to the public supply, there is a requirement to consider connection to a source of electrical energy that is not a public supply. These sources are grouped in BS7671 Chapter 55 as generating sets.

The generic term "generating sets" is used to incorporate any method of producing electricity that is connected to an installation (BS 7671 does not consider generating sets contained within equipment regulation 551 - 01 - 01). However, the two most likely to be met in practice are:

a) Rotary generators driven by a prime mover such as a combustion engine or an electric motor.

b) Static inverters.

Generating sets can be used to supply:

Permanent installations
Temporary installations
Portable equipment

When supplying permanent or temporary installations, generating sets can be:

The only means of supply.
Means of supply when the public supply is not available (standby supply).
A supply in parallel with the public supply.

Whatever the type of generator and the method of use, precautions have to be taken to ensure suitable and safe selection and erection. The Regulations in Section 551 of BS 7671 govern these requirements.

Generally the generating sets must be suitable for the intended use and the output characteristics must not be detrimental to the public supply or to any equipment connected to the generating set.

The fault level (short circuit current) that can be produced by a generating set is usually less than that obtainable from a public supply. Thus care must be taken to ensure that any protective device (fuse or circuit breaker) can be operated by the short circuit current produced by the generating set. In particular the design of the installation must ensure that when the generator set is supplying the loads all the protective devices for excess current, direct and indirect contact will operate as required by BS 7671. Also that discrimination between protective devices in series is maintained so that only the faulty circuit is disconnected.

Arising from this is the need to consult with the Supplier about the selection and erection of any generating sets that are or can be connected in parallel to the public supply Regulation 511 - 07 - 01. This consultation should also involve the generator supplier to establish the generator characteristics such as steady state, transient and sub-transient reactances. From which information can be ascertained the characteristics of protective devices that will operate correctly when the only supply is the generator.

When protection against indirect contact is not achieved by the initial design, consideration should be given to incorporating suplementary bonding or RCDs.

The use of supplementary bonding is often needed to achieve the required disconnection times for protection against indirect contact by the protective devices on the supply side of a static inverter.

Regulation 551 - 04 - 03 states that automatic disconnection of a standby generator supply must not rely on the earthing of the public supply. Thus a suitable separate earth electrode should be provided for the Generator star point. The resistance of this electrode is not specified in either BS 7430 or BS 7671. The value should be as low as possible to avoid instability of the Generating Set and will depend on soil conditions and the buried length of the electrode.

Regulation 551 - 04 - 06 requires additional 30mA RCD protection for generators which are not permanently fixed. Additionally for portable generating sets the protective conductors between separate items of equipment must be part of a cable or cord which complies with Table 54G. This requirement excludes most reduced CPC 'twin and earth' type cables.

Additionally, the ability of generating sets to supply starting currents to electric motors is usually less than that of the public supply. Care must be taken when selecting electric motors and their method of starting if they have to operate on a generating set supply. A particular example being fire pumps. The use of electronic starters that limit the current peaks during starting can be of considerable assistance in overcoming this problem.

All the general requirements of BS 7671 apply to installations to which a generating set may be connected. Section 551 governs the extra or modified requirements for the selection and erection of generating sets.

See Explanatory Illustrations D5, 6, 7 and 8 in this Handbook.

Index of Protection (IP) Code

FIRST NUMERAL		SECOND NUMERAL	
(a) Protection of persons against contact with live or moving parts inside enclosure (b) Protection of equipment against ingress of solid bodies		Protection of equipment against ingress of water	
No./SYMBOL	DEGREE OF PROTECTION	No./SYMBOL	DEGREE OF PROTECTION
0	(a) No Protection (b) No Protection	**0**	No Protection
1	(a) Protection against accidental or inadvertent contact by a large surface of the body, e.g. hand, but not against deliberate access (b) Protection against ingress of large solid objects greater than 50mm diameter	**1**	Protection against drops of water. Drops of water falling on enclosure shall have no harmful effect
2	(a) Protection against contact by standard finger (b) Protection against ingress of medium size bodies less than 12mm diameter and greater than 80mm length	**2**	Drip Proof: – Protection against drops of liquid. Drops of falling liquid shall have no harmful effect when the enclosure is tilted at any angle up to 15° from the vertical
3	(a) Protection against contact by tools, wires or suchlike more than 2.5mm thick (b) Protection against ingress of small solid bodies	**3**	Rain Proof: – Water falling as rain at any angle up to 60° from the vertical shall have no harmful effect
4	(a) As 3 above but against contact by tools, wires or the like, more than 1.00mm thick (b) Protection against ingress of small foreign bodies	**4**	Splash Proof: – Water splashed from any direction shall have no harmful effect
5	(a) Complete protection against contact (b) DUSTPROOF: – Protection against harmful deposits of dust, dust may enter but not in amount sufficient to interfere with satisfactory operation	**5**	Jet Proof: – Water projected from a nozzle from any direction (under stated conditions) shall have no harmful effect
6	(a) Complete protection against contact (b) DUST TIGHT: – Protection against ingress of dust	**6**	Watertight Equipment: – Protection against conditions on ships' decks, etc. Water from heavy seas or power jets shall not enter the enclosures under prescribed conditions
IP CODE NOTES The degree of protection is stated in form IPXX Protection against contact or ingress of water respectively is specified by replacing first or second X digit number tabled, e.g. IP2X defines an enclosure giving protection against finger contact but without any specific protection against ingress of water or liquid A letter A–D inclusive may be added to the IPXX reference to indicate the degree of protection against access to hazardous parts.		**7**	Protection Against Immersion in Water: – It shall not be possible for water to enter the enclosure under stated conditions of pressure and time
		8	Protection Against Indefinite Immersion in Water Under Specified Pressure: – It shall not be possible for water to enter the enclosure

Note: Use this table for General Guidance only – refer to BS EN60529 for full information on degrees of protection offered by enclosures

Figure C12(a)

Recommended Methods of Support for Cables, Conductors and Wiring Systems

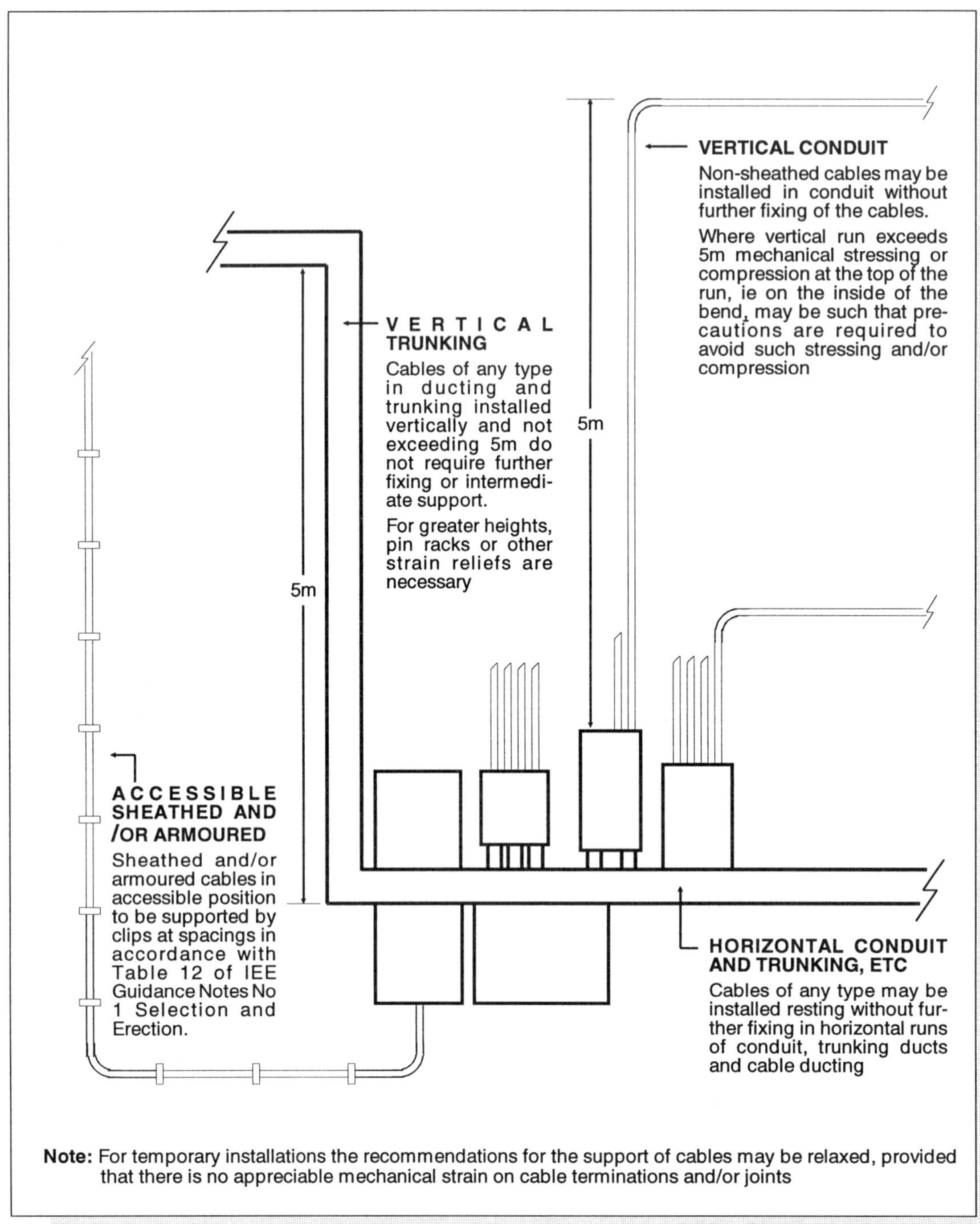

Note: For temporary installations the recommendations for the support of cables may be relaxed, provided that there is no appreciable mechanical strain on cable terminations and/or joints

Figure C12(b)

Recommended Methods of Support for Cables, Conductors and Wiring Systems (continued)

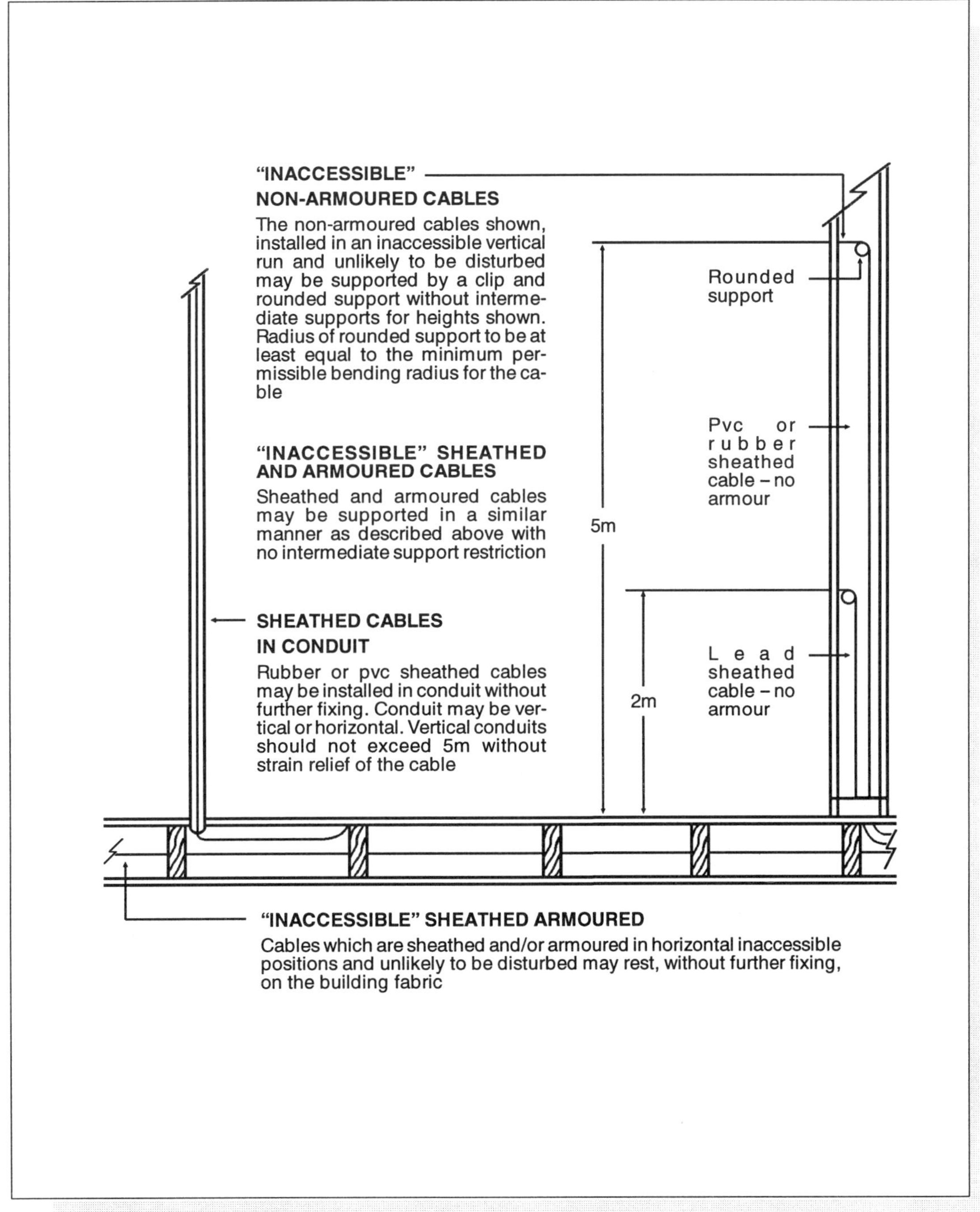

Figure C13(a)

Recommended Methods of Support for Cables used for Overhead Wiring

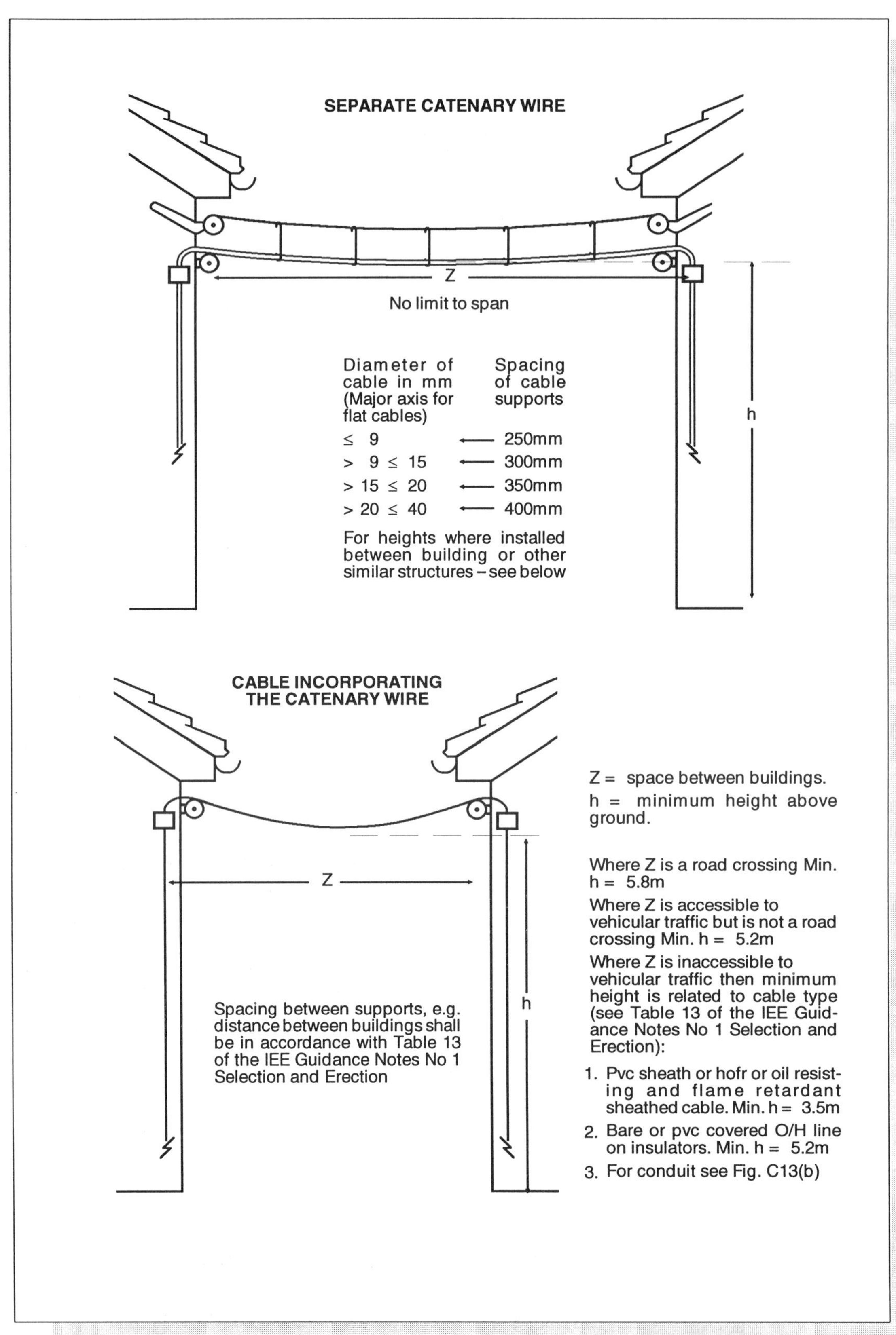

Figure C13(b)

Recommended Methods of Support for Cables used for Overhead Wiring (continued)

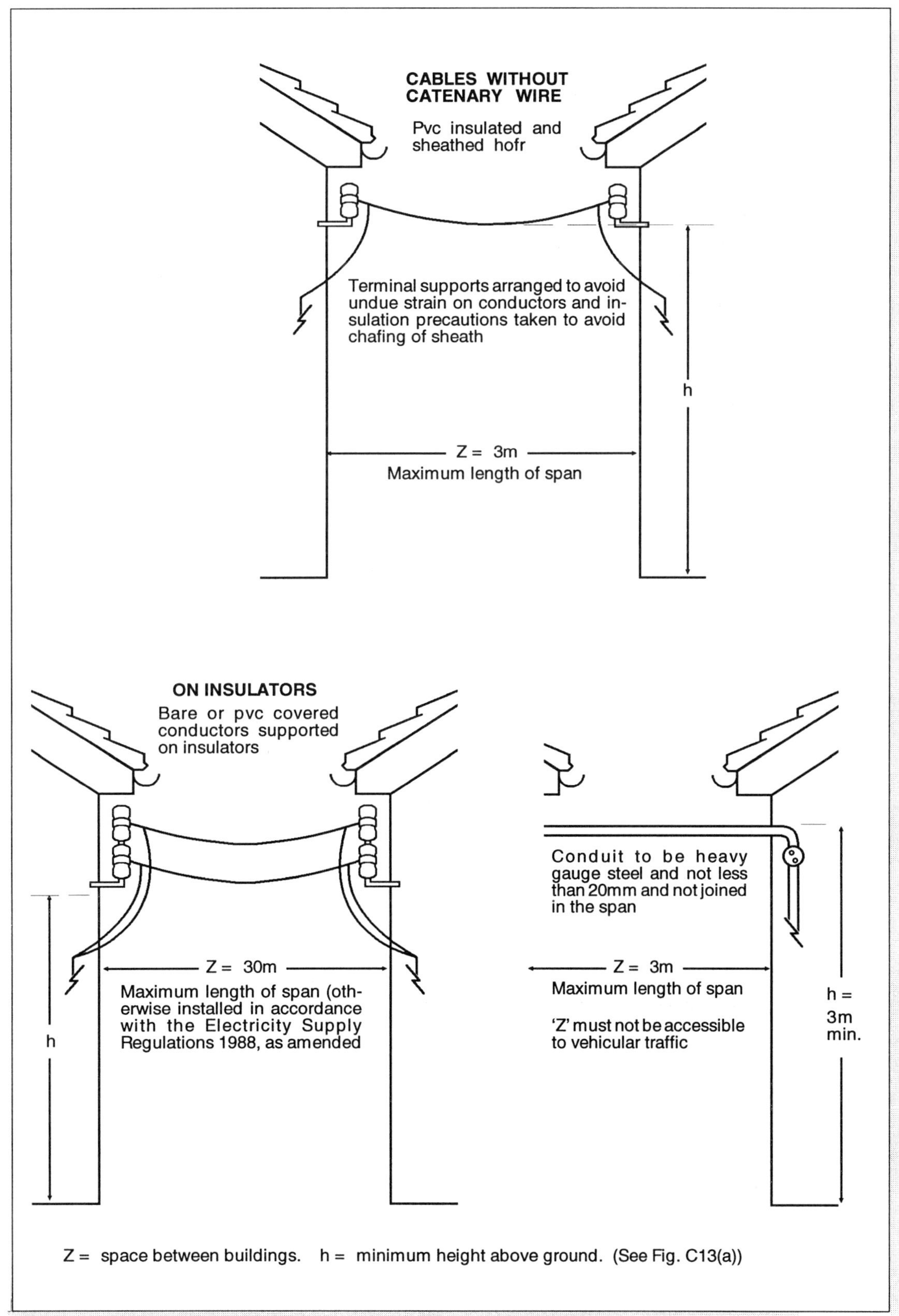

Z = space between buildings. h = minimum height above ground. (See Fig. C13(a))

Topic Chart 9

AN INFORMATION CHART

Special Installations or Locations – Particular Requirements

Electrical Installation is an assembly of associated electrical equipment supplied from a common origin to fulfil a specific purpose and having certain co-ordinated characteristics.

Restrictive Conducting Location is a location comprised mainly of metallic or conductive surrounding parts, within which it is likely that a person will come into contact through a substantial portion of his body with the conductive surrounding parts and where the possibility of preventing this contact is limited.

Highway Power Supply is an electrical installation comprising an assembly of associated highway distribution circuits, highway distribution boards and street furniture, supplied from a common origin.

Street Furniture is fixed equipment, located on a highway, the purpose of which is directly associated with the use of the Highway.

See Part 2 of BS 7671:1992 for the complete set of Definitions

TOPIC CHART 9 (Information) Special Installations or Locations – Particular Requirements

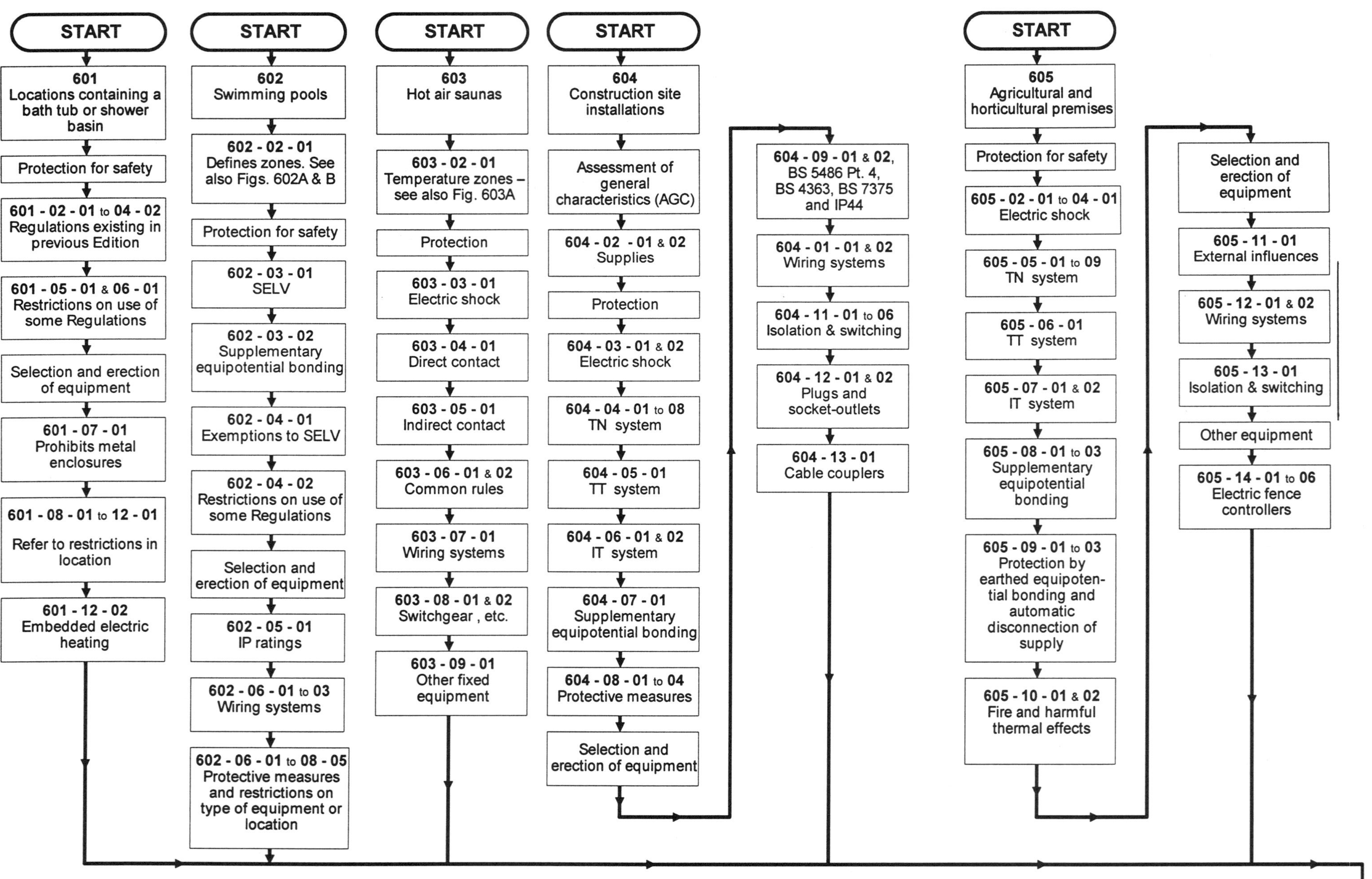

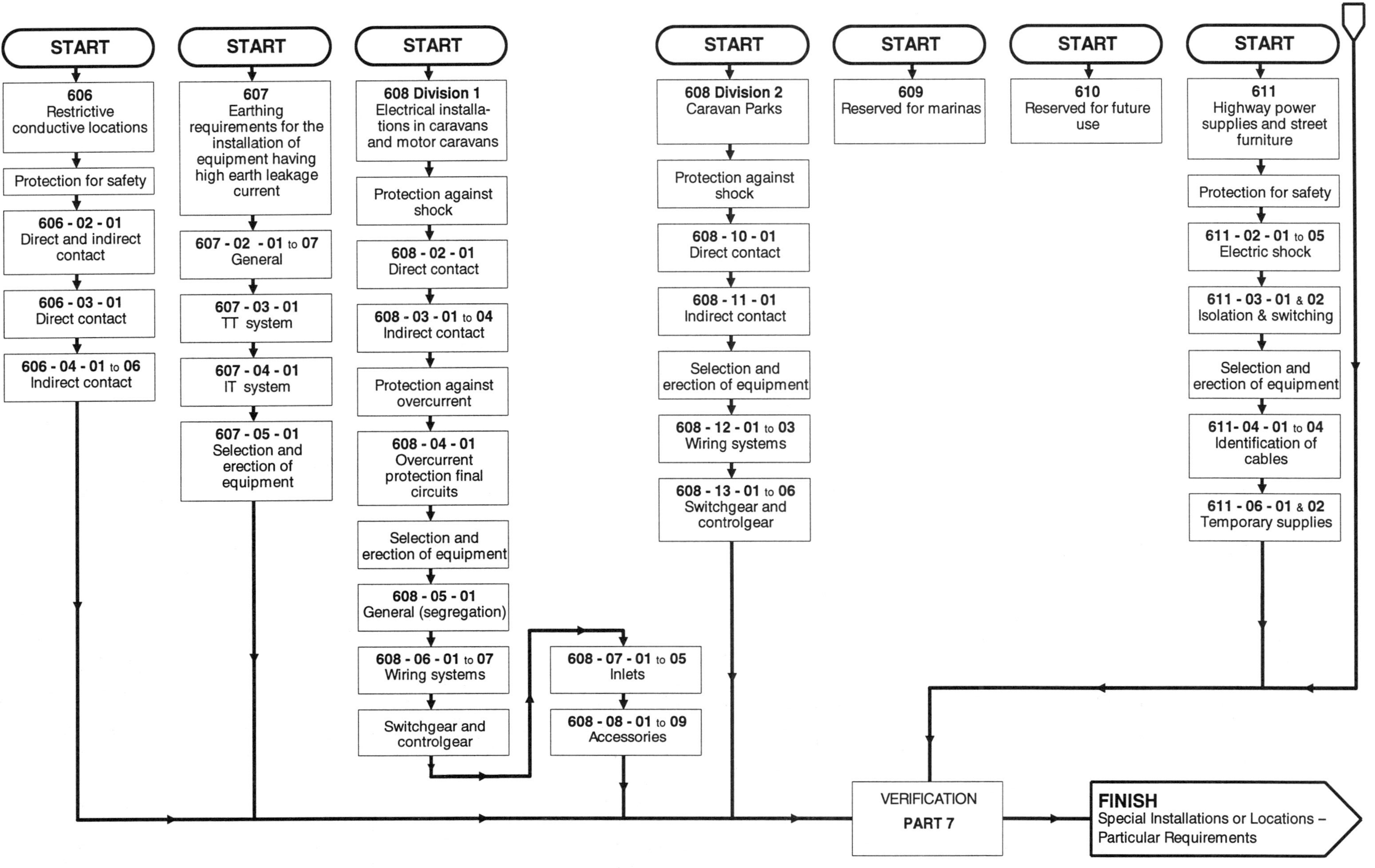

START
606 Restrictive conductive locations
Protection for safety
606 - 02 - 01 Direct and indirect contact
606 - 03 - 01 Direct contact
606 - 04 - 01 to 06 Indirect contact
START
607 Earthing requirements for the installation of equipment having high earth leakage current
607 - 02 - 01 to 07 General
607 - 03 - 01 TT system
607 - 04 - 01 IT system
607 - 05 - 01 Selection and erection of equipment
START
608 Division 1 Electrical installations in caravans and motor caravans
Protection against shock
608 - 02 - 01 Direct contact
608 - 03 - 01 to 04 Indirect contact
Protection against overcurrent
608 - 04 - 01 Overcurrent protection final circuits
Selection and erection of equipment
608 - 05 - 01 General (segregation)
608 - 06 - 01 to 07 Wiring systems
Switchgear and controlgear
608 - 07 - 01 to 05 Inlets
608 - 08 - 01 to 09 Accessories
START
608 Division 2 Caravan Parks
Protection against shock
608 - 10 - 01 Direct contact
608 - 11 - 01 Indirect contact
Selection and erection of equipment
608 - 12 - 01 to 03 Wiring systems
608 - 13 - 01 to 06 Switchgear and controlgear
START
609 Reserved for marinas
START
610 Reserved for future use
START
611 Highway power supplies and street furniture
Protection for safety
611 - 02 - 01 to 05 Electric shock
611 - 03 - 01 & 02 Isolation & switching
Selection and erection of equipment
611- 04 - 01 to 04 Identification of cables
611 - 06 - 01 & 02 Temporary supplies
VERIFICATION PART 7
FINISH Special Installations or Locations – Particular Requirements

Particular Requirements for Special Installations or Locations

The additional requirements for special installations or locations are set out in Part 6 under the following headings:

600	General
601	Locations containing a bath tub or shower basin
602	Swimming pools
603	Hot air saunas
604	Construction site installations
605	Agricultural and horticultural premises
606	Restrictive conductive locations
607	Earthing Requirements for the installation of equipment having High Earth Leakage Currents
608 Division 1	Electrical installations in Caravans and Motor Caravans
608 Division 2	Electrical Installations in Caravan Parks
609	(Reserved for Marinas)
610	(Reserved for future use)
611	Highway power supplies and street furniture

"Highway Power Supplies and Street Furniture", Section 611; this section relates to places used by the public and designated as a Highway. Similar equipment in areas used by the public which are not part of a building and which are not designated a Highway, must also comply with Section 611 requirements, e.g. pedestrian precinct or market square closed to traffic.

Figure C14

Locations Containing a Bath Tub or Shower Basin

FOR BONDING REQUIREMENTS SEE FIGURE D6

DISCONNECTION TIME

601 - 04 - 01

For circuits supplying equipment where such equipment is simultaneously accessible with exposed-conductive-parts or with other equipment or with extraneous-conductive-parts. A 0.4 second disconnection time must be achieved

STATIONARY APPLIANCE

601 - 12 - 01

No stationary appliance of which the heating elements can be touched to be installed within reach of person using bath or shower

601 - 07 - 01

No surface metallic enclosures; exposed earthing and bonding conductors

601 - 02 - 01

No electrical equipment to be installed in the interior of a bath tub or shower basin

ELECTRIC SHAVER

601 - 09 - 01

Shavers only connected through unit to BS 3535 and earth terminal of unit to be connected to its circuit protective conductor

NO SOCKET OUTLETS – except SELV

See below

601 - 10 - 02

There shall be no socket outlets and no provision for connecting portable appliances in rooms containing a fixed bath or shower (other than shaver and SELV sockets – see below)

601 - 10 - 01

12V a.c. or d.c. SELV socket outlets may be installed in bathrooms out of reach of a person using the bath if the socket has no accessible metal parts and protection against direct contact with live parts in accordance with 411 - 02 - 09

LAMPHOLDERS

601 - 11 - 01

Parts of a lampholder within a distance of 2.5m from the bath or shower shall be constructed of, or shrouded in, insulating material. Bayonet type (B22) lampholders to have shield to BS EN 61184 with a protective shield (HO Skirt) or, as an alternative, totally enclosed luminaires suitable for this type of location may be used

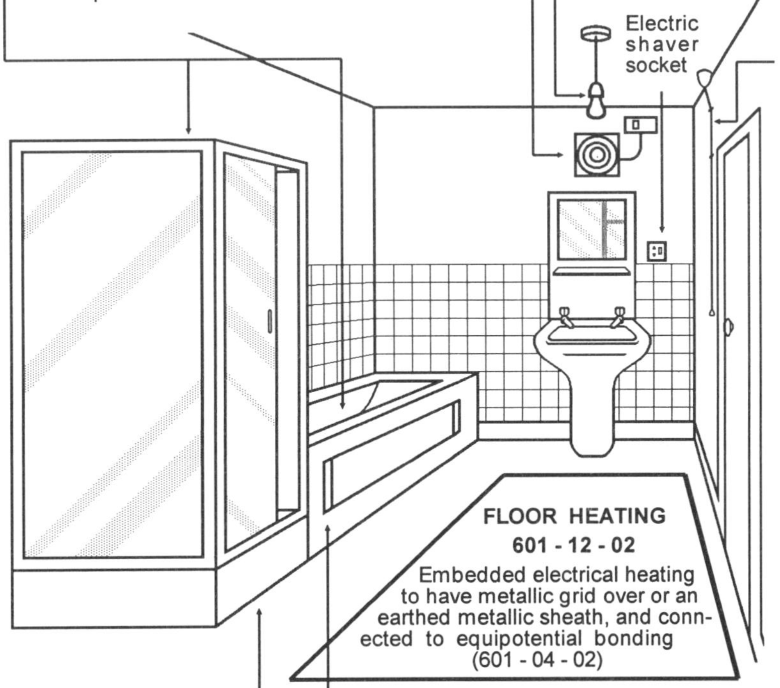

Electric shaver socket

SWITCHES AND CONTROLS

601 - 08 - 01

Every switch, electrical control or adjustment to be situated so as to be normally inaccessible to a person using the bath or shower – this does not apply to:

(i) insulating cords of switches

(ii) mechanical actuators with linkages incorporating insulating components, of remotely operated switches

*(iii) controls incorporated in water heaters and shower pumps complying with British Standards.

(iv) Switches supplied by SELV

(v) shaver supply units to comply with 601 - 09 - 01

* Notwithstanding (iii) above, it is recommended that a local cord-operated switch be provided for maintenance purposes. Cord operated switches which also provide isolation must be double pole, with adequate contact separation and clear mechanical indication of the switch position given only when both poles have achieved full separation. (Neon indicators do not suffice.) Alternatively, switches may be situated outside the bath or shower room

FLOOR HEATING

601 - 12 - 02

Embedded electrical heating to have metallic grid over or an earthed metallic sheath, and connected to equipotential bonding (601 - 04 - 02)

SHOWER CUBICLE

601 - 10 - 03

Where shower cubicles are located in rooms other than locations containing a bath tub or shower basin, any socket outlets shall be situated at least 2.5m from the cublicle

601 - 04 - 03

Bath panel to need a tool to remove it if electrical equipment installed behind

Figure C15

Caravans:
Methods of Support for Cables, Conductors and Wiring Systems
(Cables in Particular Conditions)

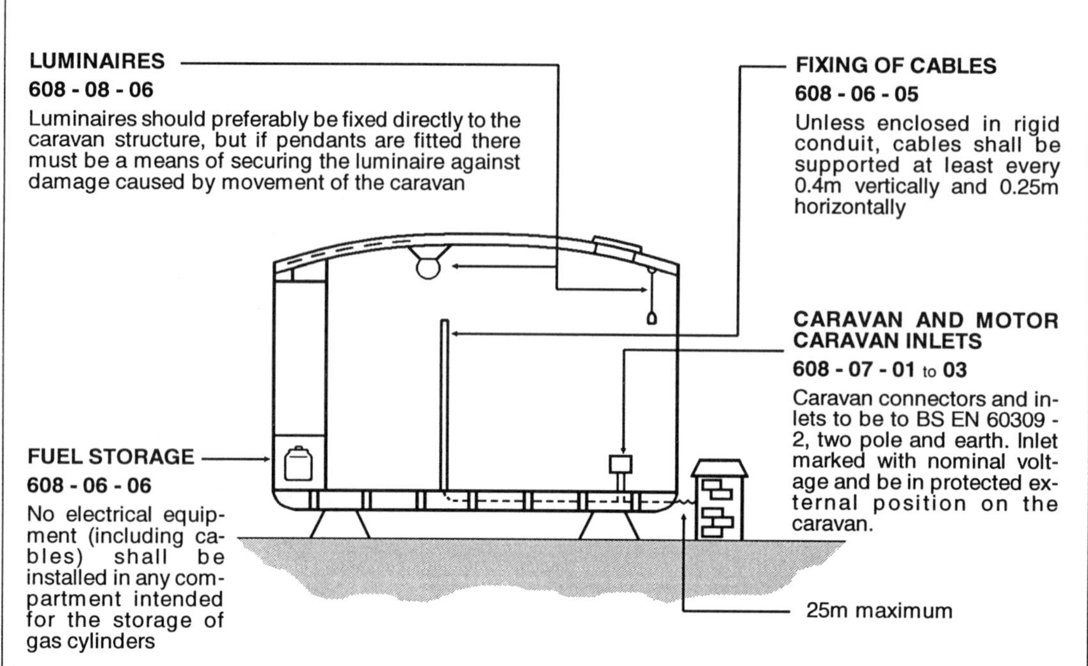

608 - 04 - 01

All final circuits to have o/c protection which disconnects all live conductors

N.B. The above relates to caravans of all types, i.e. residential and touring. Regulation 608 - 07 - 05 requires a notice to be fixed near the main switch in touring caravans. For clarity, this requirement is stated in this illustration and in Figure C15. The notice requires each rcd to be tested every time the caravan is connected to a supply.

Figure C16

Caravans, Motor Caravans, their Pitches and Caravan Parks

(For an individual maximum of 16A demand)

ALL SYSTEMS

PROTECTIVE CONDUCTORS

608 - 03 - 02

Where protection against indirect contact is afforded by automatic disconnection of supply the protective conductor to be installed throughout each circuit within the van, sockets in caravan to incorporate earthing contact

EXTRANEOUS PART BONDING

608 - 03 - 04

Extraneous-conductive-parts to be bonded to the caravan protective conductor. If construction of van does not ensure continuity, multiple bonding may be used

Minimum size of bonding conductor to be $4mm^2$

WARNING NOTICES

608 - 07 - 04

In all touring caravans, a notice (specified in this Regulation) to be fixed near the main switch inside the van. See also Regulation 514 - 12 - 02 and Figure C14

A notice concerning the periodic testing of RCDs is required at the main switchboard

CARAVAN PITCH FIXED SOCKET (a)

608 - 13 - 01 & 02

Sockets to be to BS EN 60309-2 220/230V, IPX4 minimum 16A and positioned within 20m of the caravan intake position and individually overcurrent protected

608 - 13 - 05

PME terminals are not to be connected to socket outlets supplying caravans. Protective conductors at sockets are to be connected to an independant earth electrode, i.e. to form a TT system

EARTH FAULT PROTECTION (b)

608 - 13 - 05

Each van to be supplied from a socket outlet having 30mA rcd protection. Up to 3 vans may be supplied from one rcd

TN - S SYSTEM

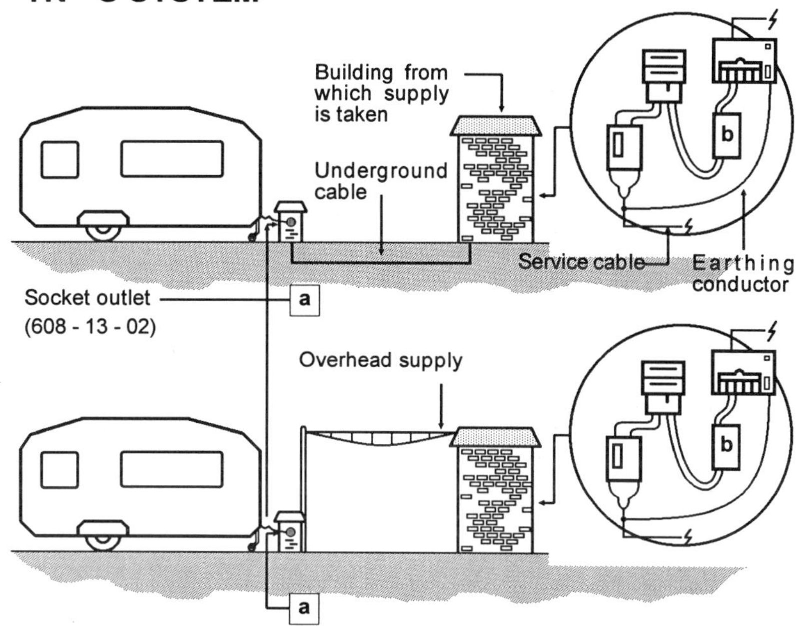

When the pitch is fed by an o.h. supply a duplicated protective conductor is required or connection to the main earthing terminal by protective conductor unlikely to fail

TT SYSTEM

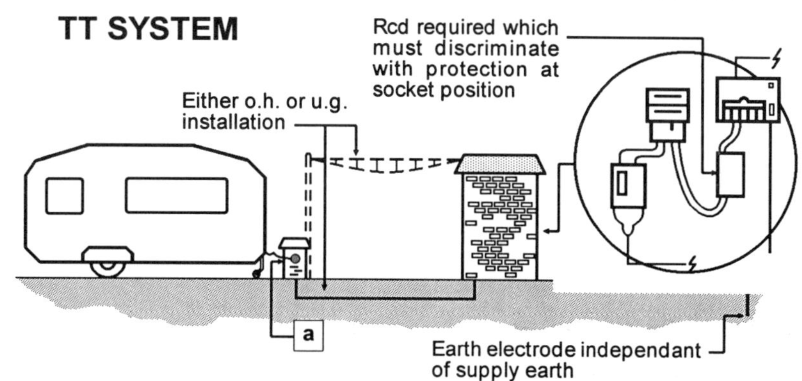

TN - C - S SYSTEM

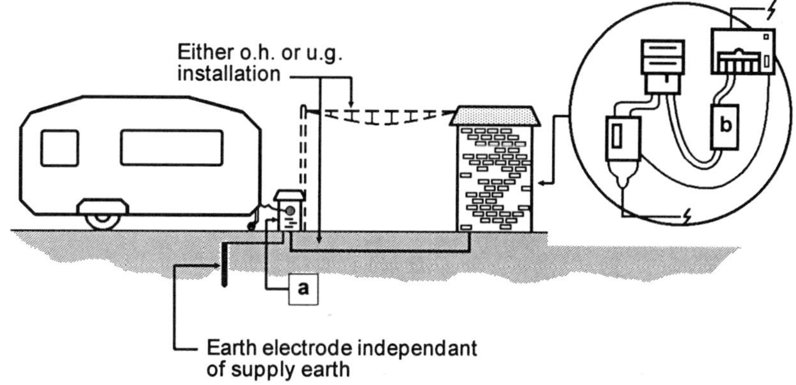

(a) Rcd and/or overcurrent protective device

(b) Position of rcd when protecting up to 3 caravans

Topic Chart 10

A DECISION CHART

Inspection and Testing

BS 7671:1992 differentiates between:

Initial verification - for new work, additions and alterations.

AND

Periodic Inspection and Testing

However the test equipment and testing methods are the same for both.

TC10 has been segregated to reflect these three separate subjects.

TC10A - Inspection and Testing for Initial Verification

Regulation 711-01-01 requires testing and inspection "during erection" to overcome some of the problems arising from having to disconnect equipment such as luminaires if testing is carried out after completion.

This Regulation also recognises that installations can only be verified so far as is "reasonably practicable" that they meet the requirements of the regulations.

On new installations certain information (see Regulations 514-09-01 and 711-01-02) must be readily available and if it is not, the requirements of BS 7671:1992 cannot be satisfied.

On complex installations it is necessary to establish at the outset the types of installation present (e.g. SELV, Functional Extra-Low Voltage, non-conducting locations, etc.) and the system employed (see page B2/2).

It is also necessary to confirm that the installation complies with Part 3 of the Regulations relating to Assessment of General Characteristics (AGC). Information regarding prospective fault current and external earth fault loop impedance will also be required.

Regulation 743 - 01 - 02 allows a Minor Electrical Works Certificate to be issued when an addition or alteration does not include the provision of any new circuit. A lighting point or socket outlet added to an existing circuit would be typical of the use of this Certificate.

TC10B - Periodic Inspection and Testing

With an existing installation, the problems may well be complicated by lack of information or by additions and alterations for which no drawings or information exist. A reasonably practicable approach must be taken to all periodic inspection and testing.

An existing installation then, should be examined to establish, among other things, whether SELV systems or extra-low voltage systems other than SELV are present. Delineation of equipotential zones should be ascertained. This may seem a simple matter, but, what may be simple to identify on a drawing, may be very difficult to appreciate on site. Equipotential zones which, on a cursory inspection, may appear to require a connection with earth, could turn out to have been designed as earth-free equipotential zones where the provision of an earth connection would, under fault conditions, lead to dangerous voltage on simultaneously accessible equipment. Furthermore,it may be that what appears to be an earth-free equipotential zone is really an equipotential zone designed to be earthed, the earth connection of which has been lost through unauthorised work or other reasons.

Clearly then, the need to establish how or if it was intended to satisfy BS 7671:1992 requirements when the installation was initially designed involves identifying what exists. Since what exists may not be all of what was originally intended, the responsibility on the person carrying out the inspection and testing is onerous.

However complex these problems may seem, they can usually be solved by a logical approach. Situations could occur, however, where there is no alternative but to accept that it is impossible to resolve the state of an installation without wholesale redesign and consequent re-construction.

TOPIC CHART 10A (Decision) Inspection and Testing for Initial Verification

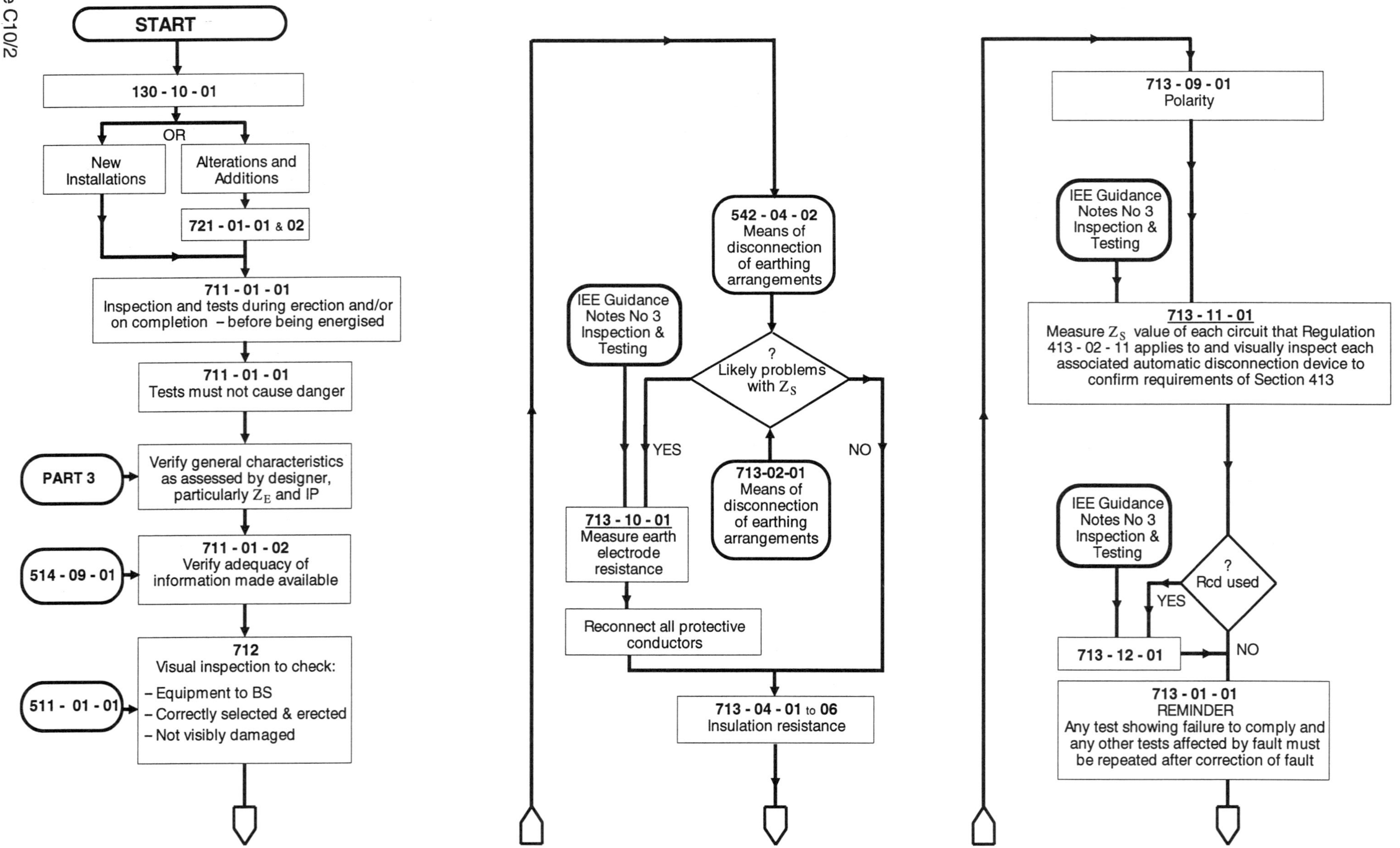

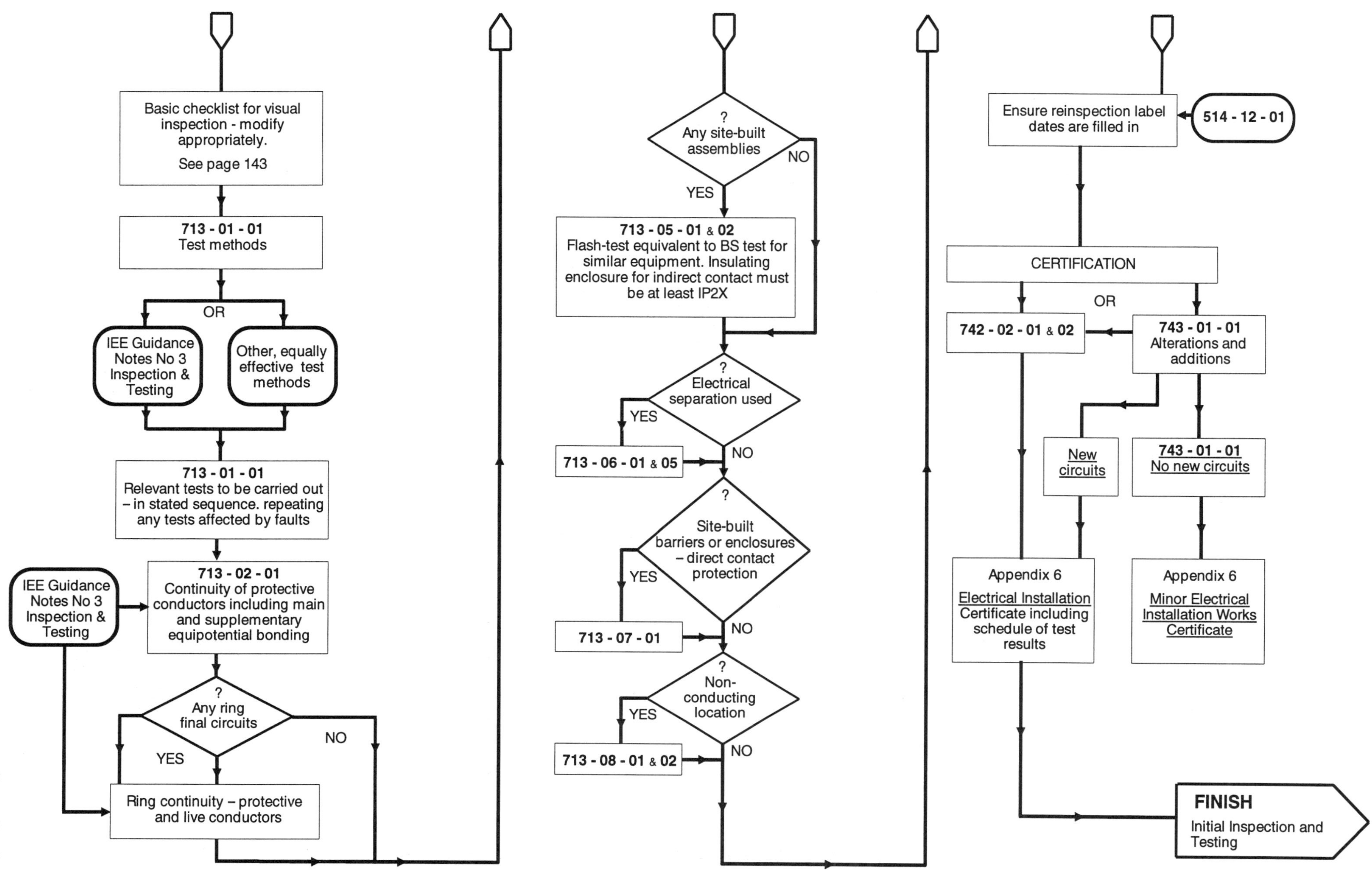
Basic checklist for visual inspection - modify appropriately.
See page 143
713 - 01 - 01
Test methods
OR
IEE Guidance Notes No 3 Inspection & Testing
Other, equally effective test methods
713 - 01 - 01
Relevant tests to be carried out – in stated sequence. repeating any tests affected by faults
IEE Guidance Notes No 3 Inspection & Testing
713 - 02 - 01
Continuity of protective conductors including main and supplementary equipotential bonding
?
Any ring final circuits
YES
NO
Ring continuity – protective and live conductors
?
Any site-built assemblies
NO
YES
713 - 05 - 01 & 02
Flash-test equivalent to BS test for similar equipment. Insulating enclosure for indirect contact must be at least IP2X
?
Electrical separation used
YES
NO
713 - 06 - 01 & 05
?
Site-built barriers or enclosures – direct contact protection
YES
NO
713 - 07 - 01
?
Non-conducting location
YES
NO
713 - 08 - 01 & 02
Ensure reinspection label dates are filled in
514 - 12 - 01
CERTIFICATION
OR
742 - 02 - 01 & 02
743 - 01 - 01
Alterations and additions
New circuits
743 - 01 - 01
No new circuits
Appendix 6
Electrical Installation Certificate including schedule of test results
Appendix 6
Minor Electrical Installation Works Certificate
FINISH
Initial Inspection and Testing

TC10C - Methods of Testing and Test Instruments

This section deals with all the requirements of test instruments for both initial verification and periodic inspection and testing.

Inspection

Inspection must be carried out before any testing is undertaken.

Regulation 712 - 01 - 03 lists the items which must be checked during the inspection. However, this list is not exhaustive, and may be supplemented or adapted for particular purposes, such as the inspection of conduit or trunking before the cables are drawn in. It is also necessary to check that the cables installed are correct relative to the nominal current and/or setting of protective devices. This list, reproduced with suitable columns in which notes or comments can be made against each item, should be available to all those responsible for inspection and testing.

The requirement that the visual inspection should include a check of the choice and setting of monitoring and protective devices (712 - 01 - 03(xii)) is a demanding responsibility because, for example, the check must include reference to the characteristics of the devices mentioned in Regulation 413 - 02 - 04, and the method used to comply with this Regulation, which will determine the required maximum earth fault loop impedance.

Obviously, the information referred to above is of considerable value to persons responsible for re-inspections and testing, and must always be noted on the Electrical Installation Certificate. It is also recommended that the Electrical Installation Certificate be retained with the record drawings and any other particulars of the installation.

Design Information

Some of the Regulations allow a degree of flexibility in the way compliance with a fundamental requirement is achieved, and in order to confirm compliance, it is necessary that those responsible for inspection and testing understand fully the design requirements and their basis. In particular, attention is drawn to the detailed information required on the Completion Certificate (BS 7671:1992 Appendix 6). The designer of the installation must provide this information. It may also be necessary to ascertain how the designed cross-sectional areas of protective conductors, etc., have been determined.

Some aspects of design are based on the prospective fault current at the origin of the installation. This value should be checked by calculation or by measurement.

Verification of Protective Devices

The sequence of measurements necessary to verify the selection and co-ordination of devices for overcurrent and shock protection is shown below.

1. Verify circuit design current (I_b)

2. Determine overcurrent protective device type and rating (I_n)

3. Determine shock protection device if different from 2. Determine operating characteristics of any rcd

4. Determine type of cable and rating factors, e.g. C_a , C_g , C_i , C_t (Appendix 4)

5. Divide I_n by correction factors = I_t

6. Determine csa of conductor having tabulated current-carrying capacity (app.4) which is the next larger than I_t

7. For verification of overcurrent protection go to Box 18

8. Determine volt-drop at furthest point of circuit

9. (i) Measure R_1 and R_2 at furthest point of the circuit applying correction factor from Appendix G of IEE Guidance Notes No 1 Selection and Erection if temperature differs from 20°C. $= \frac{(R_1 + R_2)}{C}$
   (ii) Measure Z_e
   (iii) Select multiplier M from Appendix G of IEE Guidance Notes No 1 Selection and Erection according to cable insulation type
   (iv) Determine $Z_s = Z_e + \frac{M}{C_a}(R_1 + R_2)$ (value of C_a from IEE Guidance Notes No 3)

10. Ascertain effective csa of protective conductors

11. If protective conductor effective csa satisfies Table 54G then 543 - 01 - 01 is satisfied. Go to box 14

12. If protective conductor csa does not comply with Table 54G, then the csa of the protective conductor must be established according to 543 - 01 - 03

13. For socket outlet circuits 413 - 02 - 08 is satisfied if:
   (a) Z_s (Box 9 above) is less than Table 41B1(fuses) or 41B2(mcbs), or,
   (b) Z_s (Box 9 above) is less than Table 41D, and M times R_2 is less than Table 41C value

14. For fixed equipment circuits, 413 - 02 - 09 is satisfied if Z_s (Box 9 above), is less than Table 41D value

15. Where an rcd provides shock protection, the value of Z_s (Box 9), must not exceed 50 divided by the rated operating current of the rcd
   It must be determined that the relevant disconnection times are met

16. Where an rcd provides shock protection the installation must also satisfy the requirements of 543 - 01 - 01 for minimum csa of protective conductor and the requirements for protection against thermal effects

17. Establish the operating current of the overcurrent protective device compared with lowest current-carrying capacity cable which it protects against overcurrent

18. If the design (load) current I_b is less than or equal to the nominal current I_n of the device AND:

19. If the nominal current I_n does not exceed the lowest current-carrying capacity I_t, after all the correction factors are applied then:

20. Compliance with 433 - 02 - 01 is ensured

21. Where overload device is also used for fault protection its maximum fault current breaking capacity must be compared with the prospective fault current then:

22. If breaking capacity is equal to or not less than the prospective fault current measured at the point at which the device is installed then compliance with Regulations is generally ensured with regard to fault protection for cables on load side

23. Where conditions mentioned in Box 22 do not apply it is necessary to check that the time of interruption of supply does not allow the cables to exceed the limit temperature (see 434 - 03 - 03)

24. Energy let-through of separate fault protective devices must be compared with the maximum permissible energy for the next load-side fault protective device
   Devices must be co-ordinated so that the maximum permissible let-through is not exceeded

All appendix refererences are to BS 7671:1992

TOPIC CHART 10B (Decision) Periodic Inspection and Testing

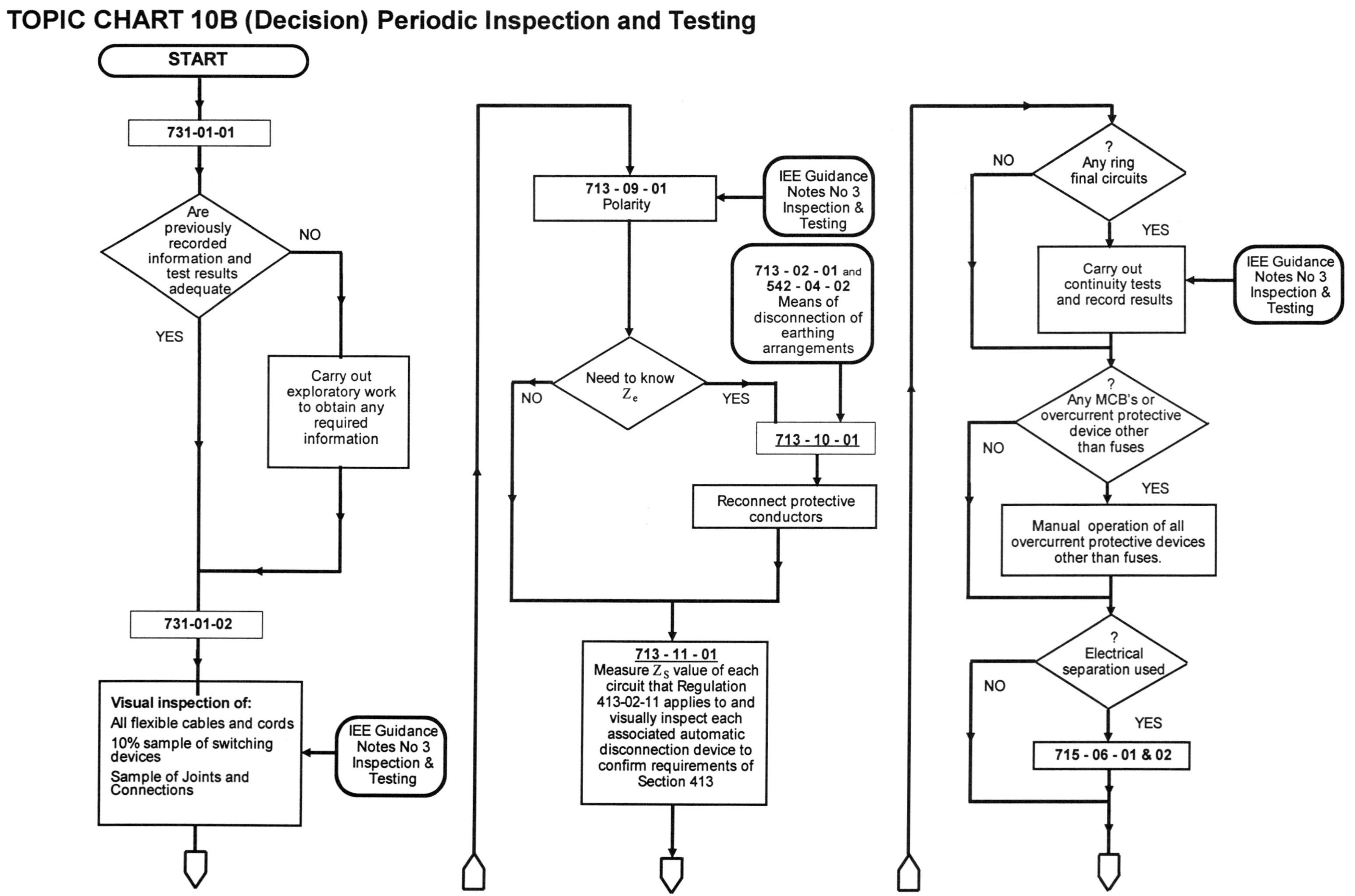

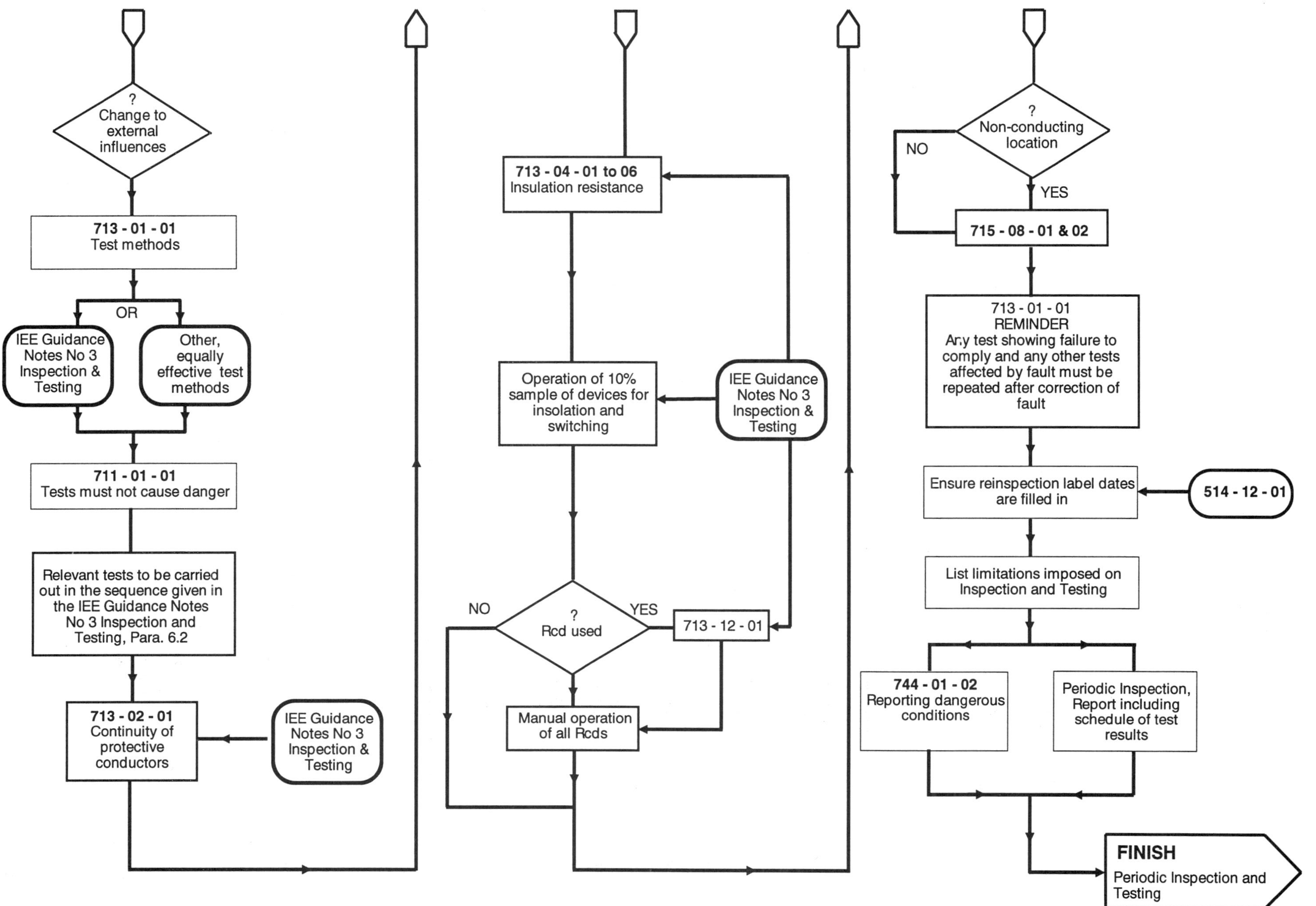

? Change to external influences
713 - 01 - 01 Test methods
OR
IEE Guidance Notes No 3 Inspection & Testing
Other, equally effective test methods
711 - 01 - 01 Tests must not cause danger
Relevant tests to be carried out in the sequence given in the IEE Guidance Notes No 3 Inspection and Testing, Para. 6.2
713 - 02 - 01 Continuity of protective conductors
IEE Guidance Notes No 3 Inspection & Testing
713 - 04 - 01 to 06 Insulation resistance
Operation of 10% sample of devices for insolation and switching
IEE Guidance Notes No 3 Inspection & Testing
NO
? Rcd used
YES
713 - 12 - 01
Manual operation of all Rcds
? Non-conducting location
NO
YES
715 - 08 - 01 & 02
713 - 01 - 01 REMINDER Any test showing failure to comply and any other tests affected by fault must be repeated after correction of fault
Ensure reinspection label dates are filled in
514 - 12 - 01
List limitations imposed on Inspection and Testing
744 - 01 - 02 Reporting dangerous conditions
Periodic Inspection, Report including schedule of test results
FINISH Periodic Inspection and Testing

Documentation and Verification

Before commencing any periodic inspection and testing the available documentation must be gathered together and assessed for compliance with the requirements of Regulation 711-01-02, also to allow testing to be carried out without danger (Regulation 711-01-01).

The Electrical Installation Certificate for the initial verification of the original installation and subsequent 'alterations and additions' together with the latest periodic inspection and schedule of test results report should be available for study and to enable a comparison of test results to be made which could indicate an incipient failure of part of the installation.

The prospective fault current at the origin of the installation may have changed due to network modifications by the Regional Electricity Company (REC). The present value should be ascertained by calculation from a measured Z_e or by measurement and compared with the recorded value. If significantly different a report on the required action should be given to the client before proceeding with the Periodic Inspection and Testing.

Minor Electrical Installation Works Certificates may be issued for small alterations that do not include the provision of a new circuit or circuits (Regulation 743 - 01 - 01). This type of certificate must be issued for each circuit altered or added to.

Any increase in load will have caused an increase in the voltage drop, which should be measured to ensure that there is still compliance with Regulations 525-01-01 and -02.

Survey

Where the documentation available does not meet the requirements given above an assessment of the shortfall must be made. The Client's agreement should be obtained to the degree of exploratory survey work, disruption to supply and cost to provide the necessary documentation to meet the requirements of future periodic inspection and test. It should be borne in mind that this work could provide much of the information required for the periodic and inspection test being undertaken.

Report

The format for the certification and reporting is set out in BS 7671:1992, see pages C10/19–C10/26. Forms for reporting and guidance on their completion based on this are available from the ECA, ECA of Scotland and the NICEIC.

For intervals between Periodic Inspections and Tests for Typical Installations (Regulation 732-01-01), see chart on page 145.

The Electrical Installation Certificate requires the interval between its issue and the first periodic inspection and testing to be inserted by the person carrying out the test for initial verification. Thereafter each Periodic Inspection report should include the recommended interval to the next Periodic Inspection. These intervals are discretionary based on experience and knowledge of the installation. However, they should not exceed the intervals shown in the following table some of which are mandatory due to statutory instruments or local authority conditions of license.

Those responsible for recommending the intervals must take into account many factors. For example, installations in kitchens of hotels and restaurants should receive careful attention, consideration being given to six-monthly tests. The changing requirements of menus, and the business generally, may lead to the addition of new equipment, which in the absence of the necessary final circuits to supply fixed equipment, may be connected to socket-outlets, with consequent overloading.

Similarly, the introduction of special office equipment may cause overloading of circuits originally intended for relatively small equipment such as typewriters, dictaphones, etc.

Much will depend on the skill and ability of those responsible for the installations, but it is recommended that contractors take care that the duties under the appropriate legislation (Electricity at Work Regulations, 1989, and Health and Safety at Work Act 1974) are brought to the attention of "the duty holder".

Under Regulations 341-01-01 and 732-01-02, it is a requirement that an assessment shall be made, so that any periodic inspection and testing can be carried out safely.

Intervals Between Periodic Inspections and Tests

Type of Installation	Reference	Period	M or R*
Domestic		10 years	R
Commercial		5 years or less	R
Agricultural and Horticultural		3 years	R
Temporary installations including construction sites		3 months	R
Caravan Parks	DOE Model Standards (1977)	1 year (3 years if underground distribution)	R
Fire Alarms	BS5839 Part 1	1 year †	R
Emergency Lighting	BS5266 Part 1	Detailed recommendations in Section 12 †	R
Cinema, etc.	Cinematograph (Safety) Regs. 1955 SI 1955 No. 1129	1 year	M
Petrol Filling Stations	By-laws under Petroleum (Regulation) Acts	1 year (usually) HS (G) 41 recommends annual	M
Launderettes	Local Authority By laws	Usually 1 year	M
Churches	"Lighting and Wiring of Churches" 1988	2 years (new installations) 1 year (installations over 5 years old)	R
Industrial	IEE Guidance Notes Inspection and Testing	3 years	R
Leisure Complexes including swimming pools	Local Authority	1 year	R
Places of public entertainment	Local Authority	1 year if G.S. 50 applies	M

*M = Mandatory   R = Recommended

† Shorter intervals are recommended for tests of operation and of batteries, generators, etc.

N.B. Items noted as "Recommended" may become "Mandatory" when incorporated in Local Authority conditions of licence for premises of various types or may be regarded as such due to the Electricity Supply Regulations 1988 (as amended) or the Electricity at Work Regulations 1989.

TOPIC CHART 10C (INFORMATION) TESTS AND TEST INSTRUMENTS

SUPPLY OFF

REGULATION NUMBER	TESTER		
	DESCRIPTION	NAME	CHARACTERISTICS
713-02 713-03	Continuity of Protective and Final Ring Circuit Conductors	Continuity Tester	O/C volts, 4 volts to 24 volts **D.C. or A.C**. Test current at least **200mA 0.01ohm** resolution 2% accuracy
713-04 & Table 71A	Insulation Resistance	Insulation Tester	Selectable D.C. output voltage of 250V, 500V, 1000V providing 1mA at the selected test voltage, 2% accuracy
713-05	Site Applied Insulation	Applied Voltage Tester	Variable voltage 0-4000V A.C. Max. output 5mA 5% accuracy
713-06	Protection by Separation of Circuits	Insulation Tester and Applied Voltage Tester	D.C. output voltage of 500V providing 1mA at the test voltage. 2% accuracy. Applied test voltage 4000V A.C. 5% accuracy
713-07	Protection by a Site Constructed Barrier or Enclosure	BS finger and Test Filament Lamp	Test voltage 40V to 50V
713-08	Insulation of Non-Conductive Floors and Walls	Insulation Tester Milliammeter Applied Voltage Tester	Test voltage 500V D.C. Providing 1mA at the test voltage. 2% accuracy Range 0 to 1mA. 2% max. accuracy with over current protection Tester voltage 2000V a.c. 5% accuracy
713-09	Polarity	ContinuityTester; Battery with Voltmeter or Bell Set	O/C volts **4 volts to 24 volts**
713-11	Earth electrode Resistance	Earth Resistance Tester	3 or 4 terminal Type. 2% accuracy

SUPPLY ON

713-09	Polarity	Voltmeter or Filament Lamp or a magnetic indicator	Class 2 BS 89
713-10	Earth fault Loop Impedance	Loop Impedance Tester	To provide approx. 25A into the test circuit, 40ms automatic cut off. 0.3 ohm to 25 ohms with 0.01 ohm resolution for up to 50A circuits
713-12	Operation of RCDs	RCD Tester	Selectable current outputs typically 10mA, 30mA, 100 mA with an accuracy 10% Tripping time indication 50 ms cut off of 5 times operating current.

Test Instruments (see also IEE Guidance Notes No 3 Inspection and Testing)

The undernoted instruments are necessary for testing to BS 7671:1992 requirements:

(1) **Continuity/resistance tester** – this should be a dual range instrument, say 0-1 and 0-10 ohms. Should measurements much below 0.05 ohm be contemplated, then consideration should be given to an instrument which passes a considerable current through the test circuit in order to verify its true value under load or fault conditions.

(2) **Earth electrode resistance tester** - to meet the requirements of BS 7430.

(3) **Insulation resistance tester** – no specific range is laid down, but a centre scale mark of 1 megohm, and a top reading (before infinity) of 50 Megohms would be generally acceptable. Test voltages of 250V, 500V and 1000V will be required.

(4) **High voltage tester** – maximum output 6kV is needed to satisfy 713-05-02(ii).

(5) **Polarity tester** – for checking before the supply is connected, a continuity tester, a battery and voltmeter or a bell-set; for live testing, a magnetic indicator, a voltmeter or a filament lamp (not a neon) with approved fused leads and test prods. N.B. Filament lamps require about 25% voltage to indicate.

(6) **Phase-earth fault impedance testers** - should have a range of at least 0.3 ohm to 25 ohms, (Regulation 413 - 02 - 10 and 11).

(7) **Rcd tester**–capable of indicating that an rcd has operated at its rated residual current, in the time specified in BS 4293; additionally, the RCD should not operate at 50% rated residual current. For a 30mA rcd providing supplementary protection against direct contact, a test current of 150mA should cause the device to operate in not more than 40ms (BS 4293).

Instruments Generally

It is recommended that instrument identification numbers be recorded against test values to avoid extensive and unnecessary re-testing when one of a number of instruments is found to be faulty. It will also help to locate the particular instrument used, where the test readings are subsequently suspect. Where an instrument has no serial number it should be marked with a unique identification.

A company Quality Assured to ISO 9000 must ensure (and for others it is strongly recommended) that all test instruments are subjected to regular checking and recalibration as necessary, and that a proper system of recording and progressing these actions is instituted. The regular use of "check boxes" or test resistors for insulation and continuity instruments is also a useful interim check.

The tests recommended for establishing the continuity of ring circuits are particularly demanding on the instruments used because of the probable low values of resistance involved, and the accuracy required. Hence, the same instrument should be used for the component parts of the testing procedure outlined on page C10/12, and great care should be taken when "shorting" test leads and adjusting the zero set control on the instrument before taking readings.

It should be remembered that meter accuracy is best in the upper half of the scale; the range giving greatest deflection of the pointer should therefore be selected. Another point to note is that no guidance is given on testing installations in hazardous atmospheres, where the instruments used for testing may themselves be a source of ignition because of the energy stored in the circuits under test. Instruments classified as intrinsically safe are only safe in relation to internal sparking within the test instrument. Precautions must be taken to ensure that the application of the test does not create a hazard.

An expectation of an infinity reading is somewhat meaningless, because a reading of infinity on one instrument design may be shown as a finite reading on an alternative instrument with a different scale or imposed voltage. For a number of mineral insulated cables tested together by an instrument of the type commonly used for insulation testing, a requirement for an infinity reading is over-optimistic. For insulation resistance measurements a value of "infinity" should never be quoted. The reading should always be expressed as being greater than the highest finite scale reading. For example, if the highest scale point is 100 Megohms, but the pointer indicates a greater value, the value should be recorded as "greater than 100 Megohms".

1. Continuity of Protective Conductors including Main and Supplementary Equipotential Bonding

Test Method 1

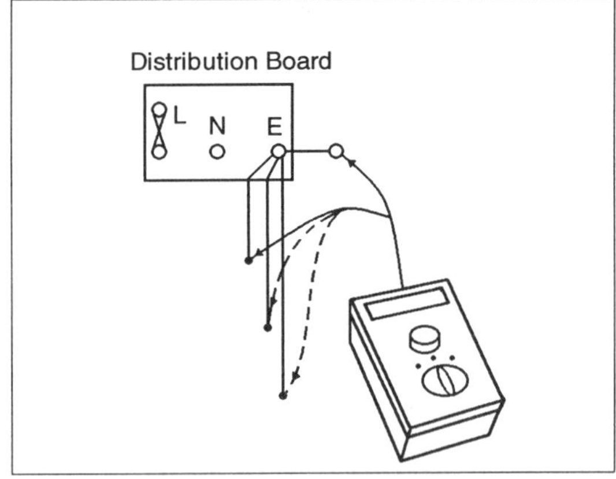

Test Method 2

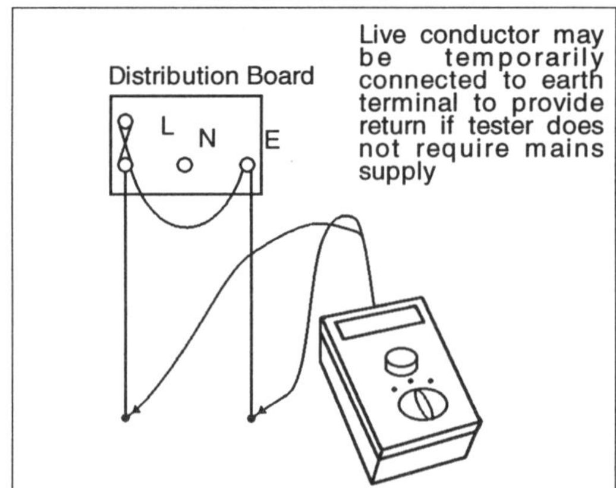

Tests may be made using either a.c. or d.c. voltages. Regulation 713 - 02 - 01 recommends tests should be made at a voltage more than 4v and no greater than 24v and a short-circuit current not less than 200mA. Where the protective conductor is not steel conduit or a steel enclosure, the requirement for the test current does not apply, and a d.c. ohmmeter may be used, e.g. when the protective conductor is copper.

Test Method (2) should only be used if the protective conductor has negligible inductance.

2. Continuity of Ring Final Circuit Conductors

The test method shown below ensures that all conductors of a ring final circuit are continuous and that no interconnecting multiple loops exist

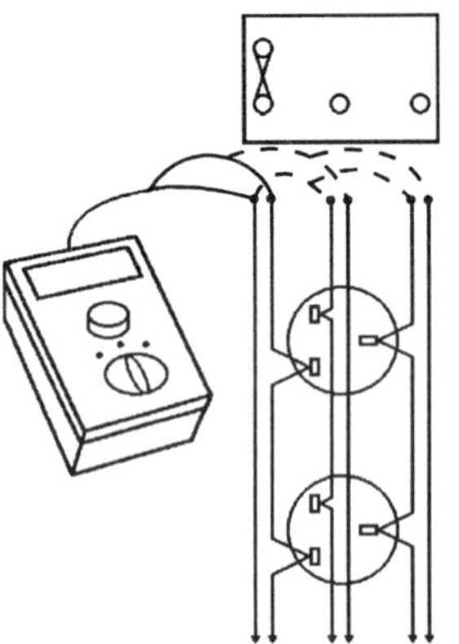

1. Initial test for continuity. With all ring conductors disconnected from the distribution board measure the resistance of the phase conductor, the neutral conductor and the cpc separately. A finite measurement indicates that the ring conductors are not open circuit.

2. Compare the measured values obtained in 1. The phase, neutral and cpc should all have the same reading ± 0.05 ohm if the cpc has the same csa as the phase and neutral conductors.

   If the cpc is a reduced size (1.5mm^2) its resistance should be approximately 1.66 times the phase conductor resistance.

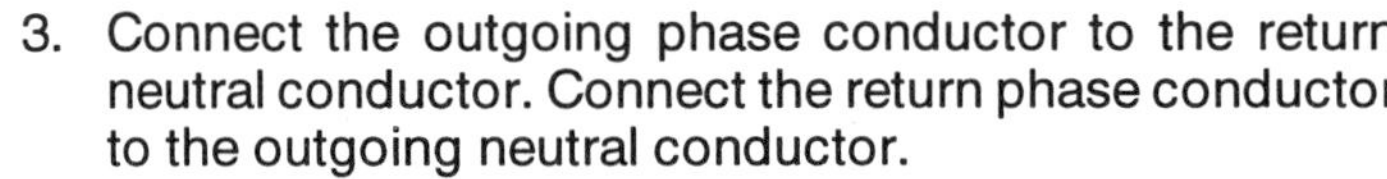

3. Connect the outgoing phase conductor to the return neutral conductor. Connect the return phase conductor to the outgoing neutral conductor.

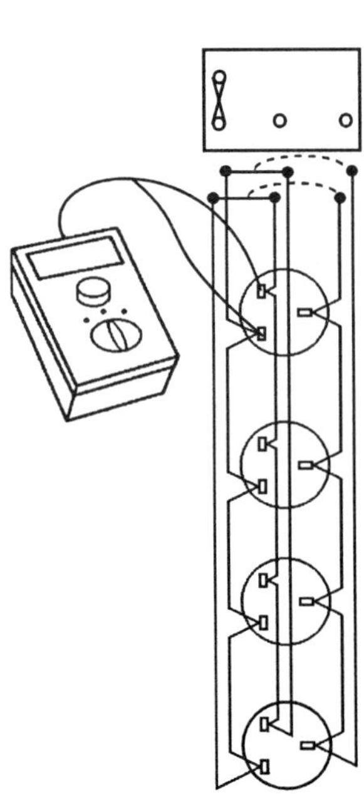

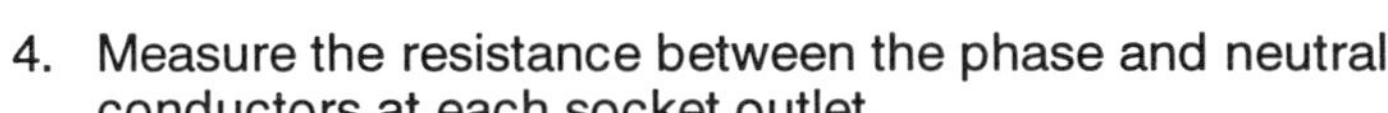

4. Measure the resistance between the phase and neutral conductors at each socket outlet.

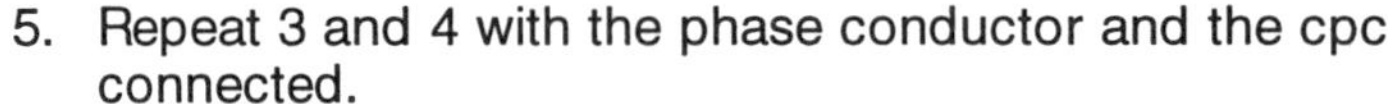

5. Repeat 3 and 4 with the phase conductor and the cpc connected.

6. All measurements at socket outlets on the ring should be approximately equal from each method of connection.

7. Socket outlets wired as a spur will have proportionately higher values dependent on the length of the spur cable.

8. The highest recorded resistance will equal $R_1 + R_2$ of the circuit and can be used to determine the earth loop impedance (Z_S) to verify compliance with BS 7671:1992 Tables 41B1 and 41B2.

9. If any significant difference is observed it will indicate a loose connection in a socket outlet or possible wrong identification of single core conductors.

False readings of continuity of ring final circuits which may occur are shown below.

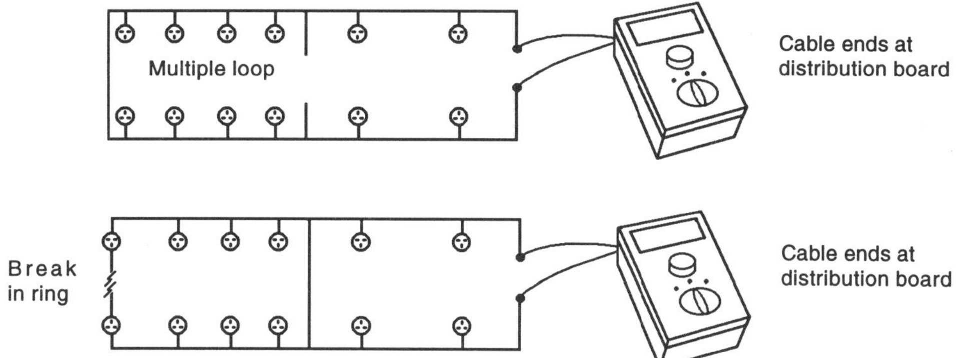

Continuity tests on ring final circuits connected as in (a) and (b) left do not prove continuity of loops, other than that nearest to the distribution board or point of test.

3. Insulation Resistance between all Conductors connected to One Phase or Pole, and in turn, all Conductors connected to each Other Phase or Pole.

Note: Any voltage sensitive equipment (e.g. control and instrumentatation equipment incorporating semi-conductor devices) must be disconnected from circuits undergoing an insulation resistance test. Failure to observe this precaution may result in serious damage necessitating costly replacements!

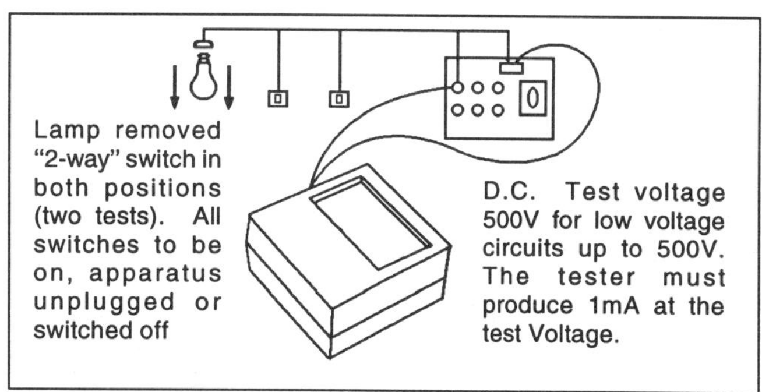

Part 7 of the BS 7671:1992 requires an insulation resistance test to be applied between poles (phase to phase and phase to neutral), and between all poles and earth, (phases and neutral to earth), using a 500V tester for installations rated up to 500V (e.g. 230/400V) and a 1000V tester for installations rated above 500V and up to 1000V.

For these tests, large installations may be divided in accordance with Regulation 713 - 04 - 02. The Regulations require the insulation resistance to be not less than 0.5 Megohm. However for a new installation a value of less than 2 Megohm warrants investigation.

Where possible, current-using equipment should be disconnected and tested separately. The insulation resistance between conductors and exposed-conductive-parts should be not less than the value given in the relevant British Standard, or where no BS exists, not less than 0.5 megohm.

With filament lamps there is no difficulty with lamp removal, followed by insulation testing between conductors and to earth. Fluorescent and other discharge lighting, may incorporate capacitor discharge resistors across the supply, so that tube removal does not enable insulation tests to be made and in addition electronic components, the presence of which may not be obvious, can be damaged by the test voltage.

It is recommended therefore that such luminaires be individually disconnected before testing final circuits.

4. Site Applied Insulation

These tests are only applicable where insulation for protection against direct contact, or supplementary insulation for protection against indirect contact (Class II equipment or equivalent insulation) is applied on site. The tests are not required where factory built and tested equipment is assembled on site.

The tests required by Regulations 713 - 05 - 01 and 02 are of a special nature, and require equipment similar to that used by manufacturers, but not usually available to those engaged in the inspection and testing of electrical installations.

For example, the applied voltage test required by Regulation 713 - 05 - 01 for insulation intended to afford protection against direct contact is required to be equivalent to that specified in the British Standard for similar factory built equipment. For switchgear and control gear, BS EN 60439 Part 1 is applicable, and where the insulation is rated at between 300V and 660V a.c., the applied voltage test is 2500V a.c. 50 Hz for equipment operating at 200V to 450V.

These tests should be regarded as being of a special nature and not normally included in the programme of testing an installation.

5. Protection by Separation of Circuits

Electrical separation is required by Regulation 411 - 02 - 05 (i) where protection against both direct and indirect contact is afforded by SELV and may also be adopted as a method of protection against indirect contact in accordance with Regulations 413 - 06 - 01 to 05.

In both the above examples, it is necessary to verify the degree of electrical separation:

(i) at the source of supply to the separated circuit,

(ii) within equipment having live parts connected to normal and separated circuits,

(iii) between conductors of normal and separated circuits.

Where the supply is derived from a safety isolating transformer complying with BS 3535, it is not necessary to verify item (i) above. In all other instances it must be verified that the source provides an equivalent degree of safety, if necessary by applying the insulation and electric strength tests specified in BS 3535 i.e. 3750 volts a.c. if the working voltage is 200 V to 450 V.

Item (ii) above must also be verified, if necessary by applying the same tests, while item (iii) should be verified as complying with Regulation 411 - 02 - 06 and 413 - 06 - 01(ii) and (iii) as appropriate.

The tests specified in BS 3535 are insulation resistance and electrical strength. The values depend on the type of the Transformer.

6. Protection against Direct Contact, by Barriers or Enclosures

Where protection against direct contact is applied on the site by the use of barriers or enclosures, tests must be carried out to verify that these measures meet the requirements of Regulations 413 - 03 - 01 to - 04. These tests are generally a visual examination combined with measurement. For full site verification of IP2X, IPXXB or IP4X special measuring gear such as a B.S. finger may be required, see the IEE Guidance Notes No. 3 Inspection and Testing and BS 5490.

Bearing in mind that these are unusual circumstances, that the use of barriers or enclosures for this purpose is generally in conjunction with a measure for protection against indirect contact, and that the tests require degrees of protection to BS 5490 (see page C8/7), they are not tests which will normally form part of the testing requirements of an installation.

7. Insulation of Non-Conducting Floors and Walls

Whilst Regulation 713 - 08 - 01 specifies that the tests shall be made at not less than three points on the surface, and specifies the relationship of those points to extraneous-conductive-parts, no indication of the test method is given. Much will depend on the area of the test electrode and the pressure applied to it, as well as the cleanliness of the surfaces in contact.

The nature of the non-conducting situation is such and the details of the test procedures so important, that these tests must be regarded as of a special character requiring the advice of those experienced in such matters, and who will probably be ultimately involved in the use and application of the locations. The distinction should be recognised between non-conducting floors which are intended to insulate, and high resistance floors which are intended to discharge safely any static charges in contact with them.

8. Polarity

Polarity testing, although commonly done when the installation is energised, may better be carried out at an earlier stage using a battery and voltmeter or a bell-set. It is not only the connection of switches to the phase conductor which is important, but also fuses and socket-outlets, as well as the centre contacts of Edison screw lampholders (where the supply neutral is earthed). Following this, visual inspection is required, noting in particular that all phase conductors are correctly identified (BS 7671:1992 Table 51A).

In general, test equipment relying on neon lamps is unsuitable because of the possibility of erroneous indications, particularly on a.c. circuits. This type of tester should only be used as a secondary or supplementary device; screwdrivers containing neon lamps are not suitable as test instruments.

9. Earth Fault Loop Impedance

Where use has to be made of protective measures which require a knowledge of the value of earth-loop impedance, Regulation 713 - 09 - 01 states that the relevant impedances shall be measured, or determined by a suitable alternative method.

During design the maximum earth fault loop impedance (Z_s) for the circuits will have been established according to Regulation 413 - 02 - 10 (fuse) or - 11 (circuit breakers) (0.4 second disconnection time) or Regulation 413 - 02 - 14 (5 second disconnection time).

If Z_s has been established using the R_1 and R_2 values from table B1 in the IEE Guidance Notes No 5 - Protection against Electric Shock then Z_s will be related to a conductor temperature of 20°C.

The value of Z_s must meet the requirement of the Regulations at the normal operating temperature of the circuit. The Z_s values obtained from on-site tests do not relate to this condition as they will probably have been taken at ambient temperature. Therefore to be certain of compliance the on-site measurement must be adjusted for the conductor resistance change due to operating conditions and ambient temperature as follows

The earth loop impedance Z_s of a circuit comprises two parts:-

Z_e – external to the circuit, measured from the origin of the circuit via the earth connection

$(R_1 + R_2)$ – the loop impedance of the phase conductor and the cpc before connection to earth, and at 20°C.

During operation, the temperature of the conductor will rise due to the current flow – which will increase the resistance. A factor M is used to correct for this effect. The value of M is dependent on the maximum temperature that the insulation is permitted to attain during operation and must be related to the type of cable being installed or tested. This correction must be used to increase the value of $(R_1 + R_2)$ from the Table B1 value, and is always a multiplier greater than 1. Factor M can be obtained from the specific cable manufacturer or from the formula on page 152.

Thus for calculation for design at 20°C:

$$Z_s = Z_e + M(R_1 + R_2).$$

When Z_s has to be related to ambient temperatures other than 20°C a correction factor C has to be applied to the resistance of the phase conductor and the resistance of the cpc. This factor is a multiplier of less than 1 for temperatures below 20°C and greater than 1 for temperatures above 20°C, thus decreasing the resistance below 20°C or increasing the resistance above 20°C.

Thus for a calculation to define a value for a design incorporating an ambient temperature that is not 20°C:

$$Z_S = Z_e + [M(R_1 + R_2)] \times C_a$$

It should be appreciated however, that the application of this correction factor only amounts to ± 2% for ± 5°C

Measurement of earth fault loop impedance may be carried out using a phase-earth loop tester as described in the IEE Guidance Notes No 3 Inspection and Testing. Neutral-earth loop testers are not recognised for this test, neither are those testers which only give a pass/fail indication because it is necessary to record actual test values.

Before performing an earth loop test, it should be checked that the earthing conductor correctly connects the installation to the means of earthing to avoid the rise in potential on exposed-conductive-parts which would otherwise occur during the test. The correctness of polarity of the supply must also be verified, and both those test facilities are normally provided in the instrument.

Alternatively, it must be ensured that no person (or livestock) can come into contact with exposed-conductive-parts for the duration of the testing, e.g. by evacuating the premises.

It should be noted that measurement of Z_e is to be made with the installation disconnected from the means of earthing. Clearly, the main switch of the installation must be in the off position during this test.

To verify compliance with the impedance values given in Tables 41B1, 41B2 and 41D, it is necessary to make two measurements, one at the origin of the installation to determine the external earth fault loop impedance Z_e and the other at the far end of each circuit to determine Z_s the total earth fault loop impedance at ambient temperature. For cables up to 35mm^2 csa, (where the inductance of the conductors is negligible compared to their resistance)

$$Z_{s\ (measured)} = Z_e + (R_1 + R_2) \qquad (1)$$

where $(R_1 + R_2)$ is related to the temperature at which the test is carried out and does not allow for the increase in resistance due to the temperature during operation.

Using the value of $(R_1 + R_2)$ from (1)

$$Z_{S\ (design)} = Z_e + [M(R_1 + R_2)\frac{1}{C_a}] \qquad (2)$$

The value of $\frac{1}{C_a}$ for various ambient temperatures can be obtained from the IEE Guidance Notes No 3 - Inspection and Testing. Typical values for M for cable types are:

Insulation type	54B Factor	54C Factor
70°C PVC	1.04	1.2
85° Rubber	1.04	1.26
90°C Thermosetting	1.04	1.28

54B applies when CPC is not part of or bunched with cables.
54C applies when CPC is incorporated in or bunched with cables

For practical reasons the measurement of Z_s is often made with the main equipotential bonding conductor in place, thus giving a lower reading than would have been obtained if the bond had been removed.

Thus it is often preferable to measure R_1 and R_2 separately

where R_1 is the resistance of the phase conductor (up to 35mm^2 csa)
R_2 is the resistance of the circuit protective conductor.

The measured values of R_1, R_2 and Z_e can then be used in the above formula to calculate $Z_{s\ (design)}$ to compare with tables 41B1, 41B2 and 41D.

The above formula and test are equally applicable for a separate cpc or where the cpc is wholly or mainly conduit or trunking.

To measure Z_s at the most distant lighting point of an installation it is essential to make the test at that ceiling rose. To plug the instrument into a convenient socket outlet and to connect the wander lead to the protective conductor at the lighting point will give low and erroneous readings.

Where the design is to Regulation 413 - 02 - 12 it is always necessary to measure the impedance of the cpc (R_2). This value must be corrected for the ambient temperature, checked against the appropriate value in Table 41C and recorded.

$$R_2 \times \frac{1}{C_a}$$

If the phase conductor is larger than 35mm^2 R_1 is not applicable and a loop impedance test must be carried out.

Where it is necessary to apply an earth fault loop impedance test to an installation or circuit protected by an rcd, it is likely that the test will trip the rcd, especially if its operating current is 30mA or less. To avoid the risks which would result from short-circuiting the rcd, it is recommended that the value should be ascertained by the use of an instrument designed for this purpose or by separate measurements of external and internal impedances.

10. Earth Electrode Resistance

The method of testing required under Regulation 713 -10 - 01 is fully described in BS 7430: 1991 (which supersedes CP 1013 – Earthing).

It is important to appreciate, however, that in this method, the electrode Z(P2) must be in a straight line, between the X(E) and Y(C2) electrodes. In built-up areas, and in certain types of ground, it may be difficult to achieve this and find suitable places to drive the electrodes.

As described in BS 7430 an alternative method is possible by making a measurement of the test electrode resistance in series with another electrode of known negligible resistance, such as a cable sheath or well-earthed water pipe. This method should only be used if the site limitations make it impossible to use the method described in BS 7430, because there will always be a doubt as to whether the resistance areas of the two electrodes are sufficiently independent and, therefore, whether an accurate result is being obtained.

11. Functional Testing

Regulation 713 - 12 - 02 requires all components of the installation to be operated to show they are properly mounted, adjusted and installed in accordance with the relevant requirements of BS 7671: 1992.

Operation of Residual Current Operated Devices

The following are general notes on residual current devices.

Passive Types. This type of rcd does not involve amplification of the residual operating current, and consequently does not require the use of electronic circuitry.

Passive types may be separated into two categories: (a) Polarised, and (b) Saturation.

As regards the Polarised category, these are field weakening devices using an armature which is retained in position by a permanent magnet. The excitation current, which is the difference between the flow and return conductor currents at any instant cancels this retaining flux by producing an opposing flux around the transformer.

When the retaining flux is cancelled, the armature is released by springs and the switch mechanism is operated.

Because the flux is affected by an alternating field, it is apparent that it is only weakened during one half of the a.c. cycle. During the other half of the a.c. cycle, the produced flux reinforces the retaining flux of the permanent magnet.

Saturation devices are those which respond to a change in flux pattern of a permanent magnet and core, due to the contributed flux resulting from an imbalance of the windings on the transformer, creating saturation in constricted areas of the laminations. This change of flux causes a weakening of the flux pattern around the armature of the transformer, which is then released by spring operation causing the mechanism to operate. Because saturation can occur during either half of the a.c. cycle, tripping will also occur during either half of the cycle.

Amplified Type. This type depends for its successful operation on the correct functioning of an amplifier, the output of which is used to operate a reed relay or other type of relay unit. As it is dependent on a supply voltage for the amplifier to operate, the device will not function if the neutral is lost and will only function within prescribed voltage tolerances. Such devices may also be damaged by the application of high d.c. voltage during testing, and the appropriate precautions should be taken.

Any residual current device provided for protection against indirect contact must be tested by means independent of the test button on the device so as to simulate an appropriate fault condition.

For a residual current device (rcd) the test requirements are given in the IEE Guidance Notes No 3 Inspection and Testing and BS 4293.

Several proprietary rcd testers are available which cause the required residual current to flow from the phase conductor on the load side of the rcd to the associated protective conductor. The test is satisfied if the rated residual current causes the rcd to open within 200 ms, or at any delay time stated by the manufacturer of the rcd.

The tester must not only trip the rcd but must also indicate the time elapsed between the start of the test and the opening of the contacts.

With a fault current flowing equivalent to 50% of the rcd rated tripping current the device should not open. With a 100% rcd rated current flowing the rcd must open in less than 200 ms. If the rcd incorporates an intentional time delay with 100% of the rated time range plus 200 ms.

Due to the variability of time delays a maximum test time cannot be specified (except in the case of the 150 mA test for supplementary protection against direct contact) however it is suggested that a 2 second maximum test time should be sufficient.

If the rcd has been installed as additional protection against direct contact its residual operating current must not exceed 30mA and the test instrument should also be capable of indicating that the rcd has opened within 40ms at a residual current of 150mA, (Regulation 412 - 06 - 02).

After applying the above tests the rcd should be tripped by means of its integral test button to check its effectiveness.

ELECTRICAL INSTALLATION CERTIFICATE

(REQUIREMENTS FOR ELECTRICAL INSTALLATIONS - BS 7671 [IEE WIRING REGULATIONS])

DETAILS OF THE CLIENT

....

INSTALLATION ADDRESS

....

....

DESCRIPTION AND EXTENT OF THE INSTALLATION Tick boxes as appropriate	
Description of installation:	New installation ☐
Extent of installation covered by this Certificate:	Addition to an existing installation ☐
(Use continuation sheet if necessary) see continuation sheet No:	Alteration to an existing installation ☐

FOR DESIGN

I/We being the person(s) responsible for the design of the electrical installation (as indicated by my/our signatures below), particulars of which are described above, have exercised reasonable skill and care when carrying out the design hereby CERTIFY that the design work for which I/we have been responsible is to the best of my/our knowledge and belief in accordance with BS 7671, amended to .... (date) except for the departures, if any, detailed as follows:

Details of departures from BS 7671 as amended (Regulations 120-02, 120-05):

The extent of liability of the signatory or the signatories is limited to the work described above as the subject of this Certificate.

For the DESIGN of the installation: **(Where there is mutual responsibility for the design)

Signature: .... Date: .... Name (IN BLOCK LETTERS): .... Designer No 1

Signature: .... Date: .... Name (IN BLOCK LETTERS): .... Designer No 2**

FOR CONSTRUCTION

I/We being the person(s) responsible for the construction of the electrical installation (as indicated by my/our signatures below), particulars of which are described above, have exercised reasonable skill and care when carrying out the construction hereby CERTIFY that the construction work for which I/we have been responsible is to the best of my/our knowledge and belief in accordance with BS 7671, amended to ....(date) except for the departures, if any, detailed as follows:

Details of departures from BS 7671, as amended (Regulations 120-02, 120-05):

The extent of liability of the signatory is limited to the work described above as the subject of this Certificate.

For CONSTRUCTION of the installation:

Signature: .... Date: .... Name (IN BLOCK LETTERS): .... Constructor

FOR INSPECTION & TESTING

I/We being the person(s) responsible for the inspection & testing of the electrical installation (as indicated by my/our signatures below), particulars of which are described above, have exercised reasonable skill and care when carrying out the inspection & testing hereby CERTIFY that the work for which I/we have been responsible is to the best of my/our knowledge and belief in accordance with BS 7671, amended to ....(date) except for the departures, if any, detailed as follows:

Details of departures from BS 7671 as amended (Regulations 120-02, 120-05):

The extent of liability of the signatory is limited to the work described above as the subject of this Certificate.

For INSPECTION AND TEST of the installation:

Signature: .... Date: .... Name (IN BLOCK LETTERS): .... Inspector

NEXT INSPECTION

I/We the designer(s), recommend that this installation is further inspected and tested after an interval of not more than .... years/months.

PARTICULARS OF SIGNATORIES TO THE ELECTRICAL INSTALLATION CERTIFICATE

Designer (No 1)

Name: .... Company: ....
Address: ....
.... Postcode: .... Tel No: ....

Designer (No 2)
(if applicable)

Name: .... Company: ....
Address: ....
.... Postcode: .... Tel No: ....

Constructor

Name: .... Company: ....
Address: ....
.... Postcode: .... Tel No: ....

Inspector

Name: .... Company: ....
Address: ....
.... Postcode: .... Tel No: ....

SUPPLY CHARACTERISTICS AND EARTHING ARRANGEMENTS Tick boxes and enter details, as appropriate

Earthing arrangements	**Number and Type of Live Conductors**		**Nature of Supply Parameters**	**Supply Protective Device Characteristics**
TN-C ☐	a.c. ☐	d.c. ☐	Nominal voltage, U/U_o (1) V	Type: ....
TN-S ☐	1-phase, 2 wire ☐	2-pole ☐	Nominal frequency, f (1) ....Hz	
TN-C-S ☐	2-phase, 3-wire ☐	3-pole ☐	Prospective fault current, I_{pf} (2) ....kA	Nominal current rating ....A
TT ☐	3-phase, 3 wire ☐	other ☐	External loop impedance, Z_e (2)......... Ω	
IT ☐	3-phase, 4 wire ☐		*(Note: (1) by enquiry, (2) by enquiry or by measurement)*	
Alternative source ☐ of supply (to be detailed on attached schedules)				

PARTICULARS OF INSTALLATION REFERRED TO IN THE CERTIFICATE Tick boxes and enter details, as appropriate

Means of Earthing

Supplier's facility ☐

Installation earth electrode ☐

Maximum Demand

Maximum demand (load) .... Amps

Details of Installation Earth Electrode (*where applicable*)

Type (e.g. rod(s), tape etc)	Location	Electrode resistance to earth
....	....	.... Ω

Main Protective Conductors

Earthing conductor: material .... csa .... connection verified ☐

Main equipotential bonding conductors material .... csa .... connection verified ☐

To incoming water and/or gas service ☐ To other elements: ....

Main Switch or Circuit-breaker

BS, Type and No. of poles .... Current rating ....A Voltage rating V

Location .... Fuse rating or setting.................A

Rated residual operating current $I_{\Delta n}$ = ... mA, and operating time of .... ms (at $I_{\Delta n}$) (applicable only where an RCD is suitable and is used as a main circuit-breaker)

COMMENTS ON EXISTING INSTALLATION (in the case of an alteration or additions see Regulation 743-01-04):
....
....
....
....

SCHEDULES
The attached Inspection and Test Result Schedules are part of this document and this Certificate is only valid when Test Result Schedules are attached to it.
.... Inspection Schedules and .... Test Result Schedules are attached.
(Enter quantities of schedules attached).

ELECTRICAL INSTALLATION CERTIFICATE
NOTES:

1. The Electrical Installation Certificate is to be used only for the initial certification of a new installation or for an alteration or addition to an existing installation where new circuits have been introduced.

   It is not to be used for a Periodic Inspection for which a Periodic Inspection Report form should be used. For an alteration or addition which does not extend to the introduction of new circuits, a Minor Electrical Installation Works Certificate may be used.

   The original Certificate is to be given to the person ordering the work (Regulation 741-01-01). A duplicate should be retained by the contractor.

2. This Certificate is only valid if accompanied by the Schedule(s) of Test Results.

3. The signatures appended are those of the persons authorised by the companies executing the work of design, construction and inspection and testing respectively. A signatory authorised to certify more than one category of work should sign in each of the appropriate places.

4. The time interval recommended before the first periodic inspection must be inserted (see IEE Guidance Note 3 for guidance).

5. The page numbers for each of the Schedules of Test Results should be indicated, together with the total number of sheets involved.

6. The maximum prospective fault current recorded should be the greater of either the short-circuit current (between the live conductors) or the earth fault current (between phase conductor(s) and an exposed-conductive-part).

7. The proposed date for the next inspection should take into consideration the frequency and quality of maintenance that the installation can reasonably be expected to receive during its intended life, and the period should be agreed between the designer, installer and other relevant parties.

ELECTRICAL INSTALLATION CERTIFICATE
GUIDANCE FOR RECIPIENTS (to be appended to the Certificate)

This safety Certificate has been issued to confirm that the electrical installation work to which it relates has been designed, constructed and inspected and tested in accordance with British Standard 7671:1992 (as amended) (The IEE Wiring Regulations).

You should have received an original Certificate and the contractor should have retained a duplicate Certificate. If you were the person ordering the work, but not the user of the installation, you should pass this Certificate, or a full copy of it including the schedules, immediately to the user.

The "original" Certificate should be retained in a safe place and be shown to any person inspecting or undertaking further work on the electrical installation in the future. If you later vacate the property, this Certificate will demonstrate to the new owner that the electrical installation complied with the requirements of British Standard 7671 at the time the Certificate was issued. The Construction (Design and Management) Regulations require that for a project covered by those Regulations, a copy of this Certificate and any schedules are included in the project health and safety documentation.

For safety reasons, the electrical installation will need to be inspected at appropriate intervals by a competent person. The maximum time interval recommended before the next inspection is stated on Page 1 under "Next Inspection".

This Certificate is intended to be issued only for a new electrical installation or for new work associated with an alteration or addition to an existing installation. It should not have been issued for the inspection of an existing electrical installation. A "Periodic Inspection Report" should be issued for such a periodic inspection.

MINOR ELECTRICAL INSTALLATION WORKS CERTIFICATE

(REQUIREMENTS FOR ELECTRICAL INSTALLATIONS - BS 7671 [IEE WIRING REGULATIONS])
To be used only for minor electrical work which does not include the provision of a new circuit

PART 1 : Description of minor works

1. Description of the minor works
2. Location/Address
3. Date minor works completed
4. Details of departures, if any, from BS 7671 : 1992 (as amended)

PART 2 : Installation details

1. System earthing arrangement (where known) TN-C-S ☐ TN-S ☐ TT ☐
2. Method of protection against indirect contact
3. Protective device for the modified circuit Type .... Rating .... A

Comments on existing installation, including adequacy of earthing and bonding arrangements : (see Regulation 130-09)

PART 3 : Essential Tests

Earth continuity satisfactory ☐

Insulation resistance:

Phase/neutral .... MΩ

Phase/earth .... MΩ

Neutral/earth...................................... MΩ

Earth fault loop impedance .... Ω

Polarity satisfactory ☐

RCD operation (if applicable). Rated residual operating current $I_{\Delta n}$ ....mA and operating time of ....ms (at $I_{\Delta n}$)

PART 4 : Declaration

I/We CERTIFY that the said works do not impair the safety of the existing installation, that the said works have been designed, constructed, inspected and tested in accordance with BS 7671 : 1992 (IEE Wiring Regulations), amended to .... and that the said works, to the best of my/our knowledge and belief, at the time of my/our inspection, complied with BS 7671 : 1992 except as detailed in Part 2.

Name:	Signature: ....
For and behalf of:	Position: ....
Address:	
....	Date : ....
....	

MINOR ELECTRICAL INSTALLATION WORKS CERTIFICATE
GUIDANCE FOR RECIPIENTS (to be appended to the Certificate)

This Certificate has been issued to confirm that the electrical installation work to which it relates has been designed, constructed and inspected and tested in accordance with British Standard 7671: 1992 (as amended), (The IEE Wiring Regulations.)

You should have received an original Certificate and the contractor should have retained a duplicate. If you were the person ordering the work, but not the owner of the installation, you should pass this Certificate, or a copy of it, to the owner.

The Minor Works Certificate is only to be used for additions, alterations or replacements to an installation that do not extend to the provision of a new circuit. Examples include the addition of a socket-outlet or lighting point to an existing circuit, or the replacement or relocation of a light switch. A separate Certificate should have been received for each existing circuit on which minor works have been carried out. This Certificate is not valid if you requested the contractor to undertake more extensive installation work. An Electrical Installation Certificate would be required in such circumstances.

The "original" Certificate should be retained in a safe place and be shown to any person inspecting or undertaking further work on the electrical installation in the future. If you later vacate the property, this Certificate will demonstrate to the new owner that the minor electrical installation work carried out complied with the requirements of British Standard 7671 at the time the Certificate was issued.

PERIODIC INSPECTION REPORT FOR AN ELECTRICAL INSTALLATION

(REQUIREMENTS FOR ELECTRICAL INSTALLATIONS - BS 7671 [IEE WIRING REGULATIONS])

DETAILS OF THE CLIENT

Client: ....

Address: ....

Purpose for which this Report is required: ....

DETAILS OF THE INSTALLATION Tick boxes as appropriate

Occupier: ....

Installation: ....

Address: ....

Description of Premises: Domestic ☐ Commercial ☐ Industrial ☐ Other ☐

Estimated age of the Electrical Installation: .... years

Evidence of Alterations or Additions: Yes ☐ No ☐ Not apparent ☐

If "Yes", estimate age: .... years

Date of last inspection: .... Records available Yes ☐ No ☐

EXTENT AND LIMITATIONS OF THE INSPECTION

Extent of electrical installation covered by this report: ....

....

....

Limitations: ....

....

....

This inspection has been carried out in accordance with BS 7671: 1992 (IEE Wiring Regulations), as amended. Cables concealed within trunking and conduits, or cables and conduits concealed under floors, in roof spaces and generally within the fabric of the building or underground have not been inspected.

NEXT INSPECTION

I/We recommend that this installation is further inspected and tested after an interval of not more than .... months/years, provided that any observations 'requiring urgent attention' are attended to without delay.

DECLARATION

INSPECTED AND TESTED BY

Name: .... Signature: ....

For and on behalf of: .... Position: ....

Address: ....

.... Date: ....

....

SUPPLY CHARACTERISTICS AND EARTHING ARRANGEMENTS Tick boxes and enter details, as appropriate

Earthing arrangements	Number and Type of Live Conductors	Nature of Supply Parameters	Supply Protective Device Characteristics
TN-C ☐	a.c. ☐ d.c. ☐	Nominal voltage, U/U_o [(1)] V	Type: ....
TN-S ☐	1-phase, 2 wire ☐ 2-pole ☐	Nominal frequency, f [(1)] .... Hz	
TN-C-S ☐	2-phase, 3-wire ☐ 3-pole ☐	Prospective fault current, I_{pf} [(2)] .... kA	Nominal current rating ....A
TT ☐	3-phase, 3 wire ☐ other ☐	External loop impedance, Z_e [(2)]....... Ω	
IT ☐	3-phase, 4 wire ☐	(*Note: (1) by enquiry, (2) by enquiry or by measurement*)	

PARTICULARS OF INSTALLATION REFERRED TO IN THE REPORT Tick boxes and enter details, as appropriate

Means of Earthing
Supplier's facility ☐
Installation earth electrode ☐

Details of Installation Earth Electrode (*where applicable*)

Type (e.g. rod(s), tape etc)	Location	Electrode resistance to earth
....	....	.... Ω

Main Protective Conductors

Earthing conductor: material .... csa ....

Main equipotential bonding conductors material .... csa ....

To incoming water service ☐ To incoming gas service ☐ To incoming oil service ☐ To structural steel ☐
To lightning protection ☐ To other incoming service(s) ☐ (state details...)

Main Switch or Circuit-breaker

BS, Type and number of poles .... Current rating ....A Voltage rating V

Location .... Fuse rating or setting................A

Rated residual operating current $I_{\Delta n}$ = ... mA, and operating time of .... ms (at $I_{\Delta n}$) (applicable only where an RCD is suitable and is used as a main circuit-breaker)

OBSERVATIONS AND RECOMMENDATIONS Tick boxes as appropriate

Recommendations as detailed below

Referring to the attached Schedule(s) of Inspection and Test Results, and subject to the limitations specified at the Extent and Limitations of the Inspection section

☐ No remedial work is required ☐ The following observations are made:

....
....
....
....
....
....
....

One of the following numbers, as appropriate, is to be allocated to each of the observations made above to indicate to the person(s) responsible for the installation the action recommended.

1 requires urgent attention 2 requires improvement 3 requires further investigation

4 does not comply with BS 7671: 1992 (as amended). This does not imply that the electrical installation inspected is unsafe.

SUMMARY OF THE INSPECTION

Date(s) of the inspection: ....

General condition of the installation: ....
....
....
....

Overall assessment: Satisfactory/Unsatisfactory

SCHEDULE(S)

The attached Inspection and Test Result Schedules are part of this document and this Report is only valid when Test Result Schedules are attached to it.

.... Inspection Schedules and .... Test Result Schedules are attached.
(Enter quantities of schedules attached).

PERIODIC INSPECTION REPORT
NOTES:

1. This Periodic Inspection Report form shall only be used for the reporting on the condition of an existing installation.

2. The Report, normally comprising at least four pages, shall include schedules of both the inspection and the test results. Additional sheets of test results may be necessary for other than a simple installation. The page numbers of each sheet shall be indicated, together with the total number of sheets involved.

3. The intended purpose of the Periodic Inspection Report shall be identified, together with the recipient's details in the appropriate boxes.

4. The maximum prospective fault current recorded should be the greater of either the short-circuit current (between the live conductors) or the earth fault current (between phase conductor(s) and an exposed-conductive-part).

5. The 'Extent and Limitations' box shall fully identify the elements of the installation that are covered by the report and those that are not; this aspect having been agreed with the client and other interested parties before the inspection and testing is carried out.

6. The recommendation(s), if any, shall be categorised using the numbered coding 1-4 as appropriate.

7. The 'Summary of the Inspection' box shall clearly identify the condition of the installation in terms of safety.

8. Where the periodic inspection and testing has resulted in a satisfactory overall assessment, the time interval for the next periodic inspection and testing shall be given. The IEE Guidance Note 3 provides guidance on the maximum interval between inspections for various types of buildings. If the inspection and test reveals that parts of the installation require urgent attention, it would be appropriate to state an earlier re-inspection date having due regard to the degree of urgency and extent of the necessary remedial work.

9. If the space available on the model form for information on recommendations is insufficient, additional pages shall be provided as necessary.

PERIODIC INSPECTION REPORT
GUIDANCE FOR RECIPIENTS (to be appended to the Report)

This Periodic Inspection Report form is intended for reporting on the condition of an existing electrical installation.

You should have received an original Report and the contractor should have retained a duplicate. If you were the person ordering this Report, but not the owner of the installation, you should pass this Report, or a copy of it, immediately to the owner.

The original Report is to be retained in a safe place and be shown to any person inspecting or undertaking work on the electrical installation in the future. If you later vacate the property, this Report will provide the new owner with details of the condition of the electrical installation at the time the Report was issued.

The 'Extent and Limitations' box should fully identify the extent of the installation covered by this Report and any limitations on the inspection and tests. The contractor should have agreed these aspects with you and with any other interested parties (Licensing Authority, Insurance Company, Building Society etc) before the inspection was carried out.

The Report will usually contain a list of recommended actions necessary to bring the installation up to the current standard. **For items classified as 'requires urgent attention', the safety of those using the installation may be at risk,** and it is recommended that a competent person undertakes the necessary remedial work without delay.

For safety reasons, the electrical installation will need to be re-inspected at appropriate intervals by a competent person. The maximum time interval recommended before the next inspection is stated in the Report under 'Next Inspection.'

D

Explanatory Illustrations

Explanatory Illustrations

In the Regulations there is much that is not easily described in the form of commentary.

The purpose of the illustrations in this part of the Handbook is to present those requirements visually with full explanatory notes on each.

The illustrations (Figures) contained in the following pages are:

The following abbreviations are used in Figure D1 opposite

R.C. Ring Circuit
S.O. Socket-outlet (BS1363)
S.O.S. As above (with switch)
F.C.U. Fused Connection Unit (max. 13A fuse)
F.C.U.S. As above (with switch)
F.S. Fused Spur
N.F.S. Non-Fused Spur
J.B. Joint Box
S.F.U. Switch & Fuse Unit (13A max.)
C.B. Circuit Breaker (16A max.)

Figure D1

Recommended Arrangement of Domestic Ring Circuits Using Socket-outlets to BS 1363

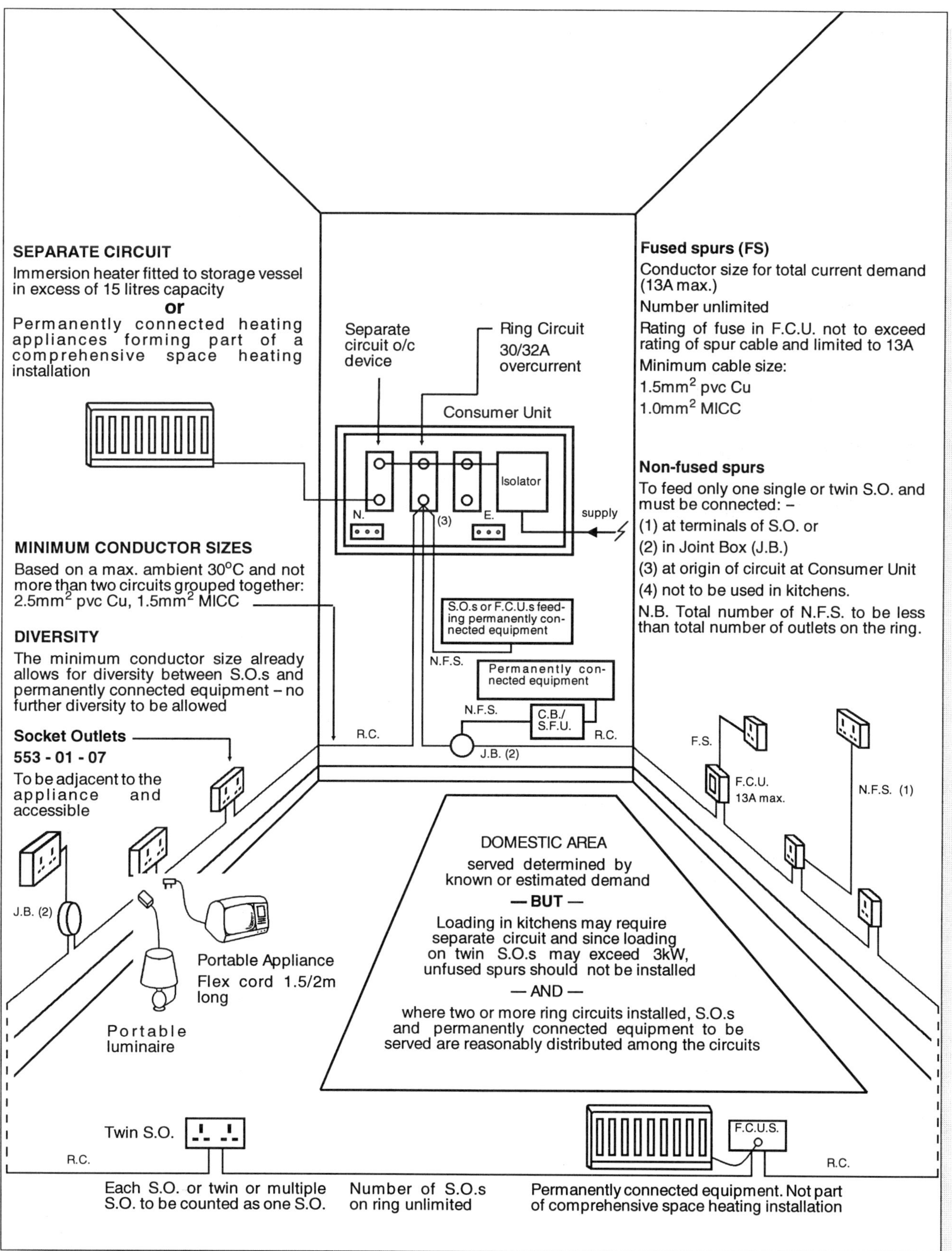

Figure D2

Recommended Arrangements of Radial Circuits, 16A Socket-outlets to BS 4343 1PH & 3PH, 110/230/415V – 50/60Hz

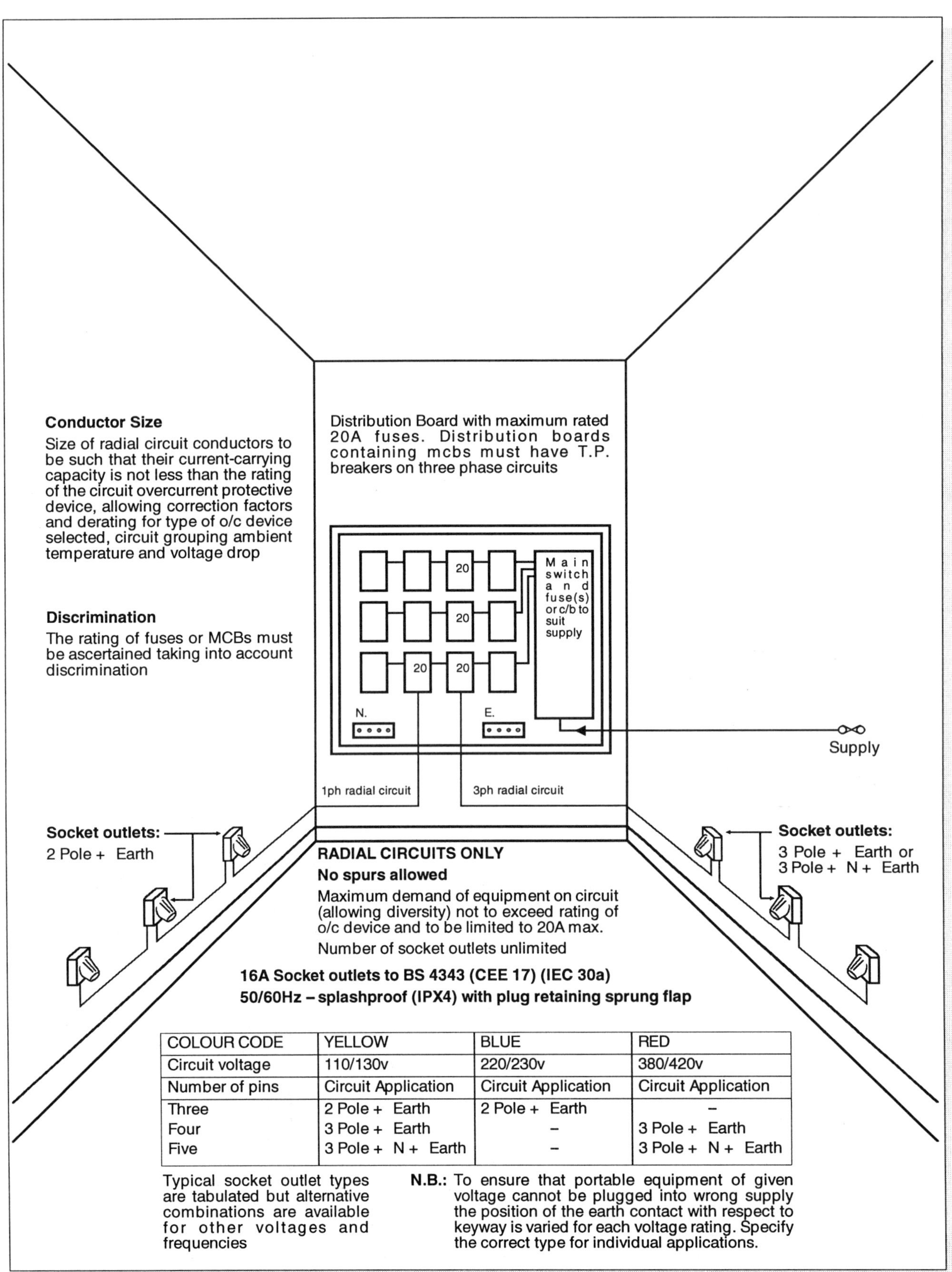

COLOUR CODE	YELLOW	BLUE	RED
Circuit voltage	110/130v	220/230v	380/420v
Number of pins	Circuit Application	Circuit Application	Circuit Application
Three	2 Pole + Earth	2 Pole + Earth	–
Four	3 Pole + Earth	–	3 Pole + Earth
Five	3 Pole + N + Earth	–	3 Pole + N + Earth

Typical socket outlet types are tabulated but alternative combinations are available for other voltages and frequencies

N.B.: To ensure that portable equipment of given voltage cannot be plugged into wrong supply the position of the earth contact with respect to keyway is varied for each voltage rating. Specify the correct type for individual applications.

Figure D3

Recommendations for Cooker Circuits

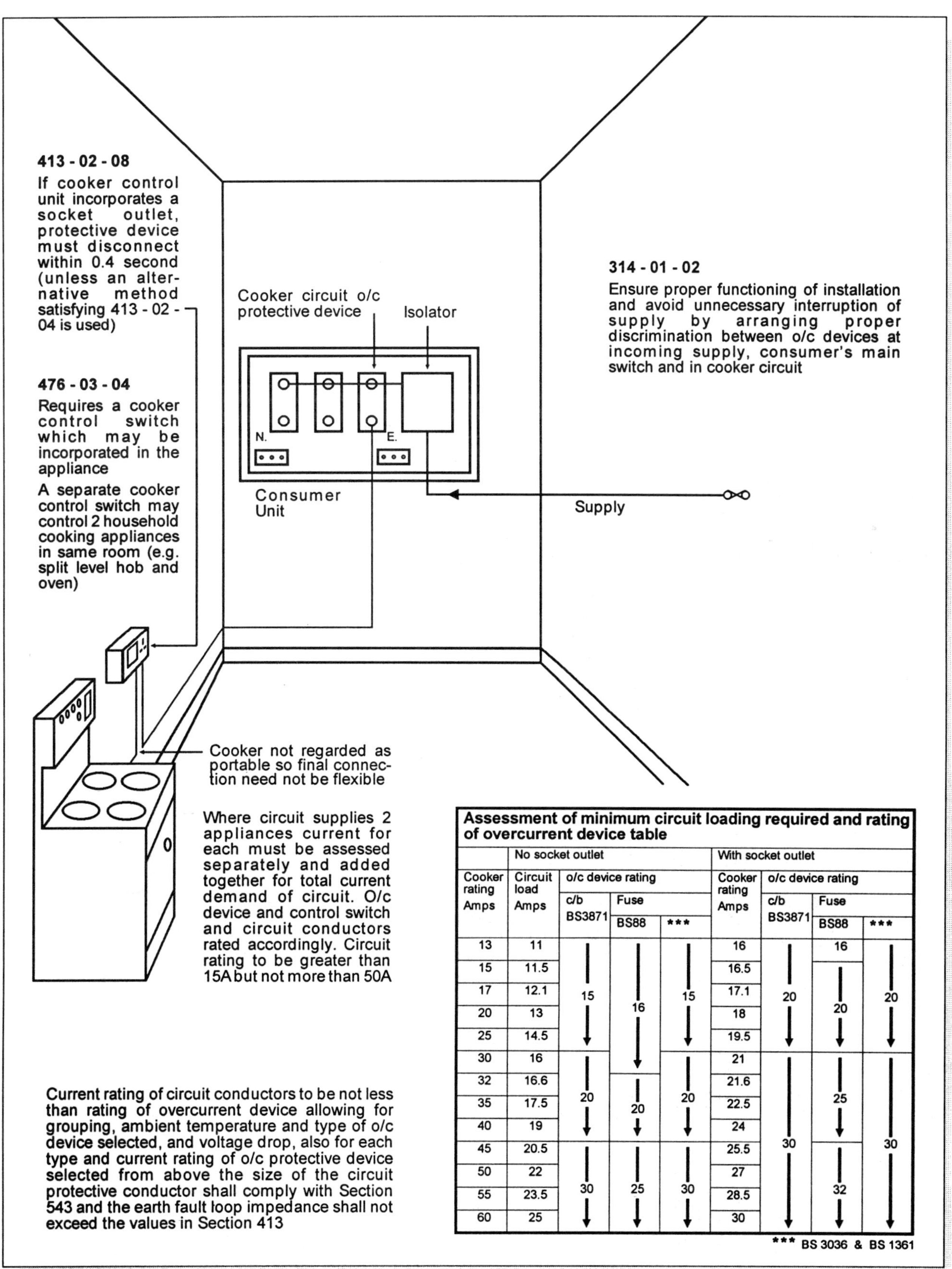

Assessment of minimum circuit loading required and rating of overcurrent device table

No socket outlet					With socket outlet			
Cooker rating Amps	Circuit load Amps	o/c device rating c/b BS3871	o/c device rating Fuse BS88	o/c device rating Fuse ***	Cooker rating Amps	o/c device rating c/b BS3871	o/c device rating Fuse BS88	o/c device rating Fuse ***
13	11				16		16	
15	11.5				16.5			
17	12.1	15		15	17.1	20		20
20	13		16		18		20	
25	14.5				19.5			
30	16				21			
32	16.6				21.6			
35	17.5	20	20	20	22.5		25	
40	19				24			
45	20.5				25.5	30		30
50	22				27			
55	23.5	30	25	30	28.5		32	
60	25				30			

*** BS 3036 & BS 1361

Figure D4

Typical Requirements for Bonding in Locations Containing a Bath Tub or Shower Basin

Note:
In the illustration the only accessible extraneous-conductive-parts are the towel rail, taps and the bath or wash basin if metal.

The illustration shows the towel rail to be water heated, i.e. an extraneous-conductive-part thus:

547 - 03 - 02

Requires the supplementary bond to be of a conductance not less than half that of the protective conductor connected to the expose- conductive-part (the heater) if the bond is mechanically protected, and 4mm^2 minimum if no mechanical protection is provided.

If the towel rail is electrically heated it is an exposed-conductive-part thus:

547 - 03 - 01

Requires the supplementary bond to be of a conductance not less than half that of the smallest protective conductor connected to the exposed-conductive-part if the bond is mechanically protected, and 4mm^2 minimum if no mechanical protection is provided

547 - 03 - 05

Where a fixed appliance such as a fan heater is supplied by a short length of flexible cord the cpc within the cord can be deemed to satisfy the bonding requirement

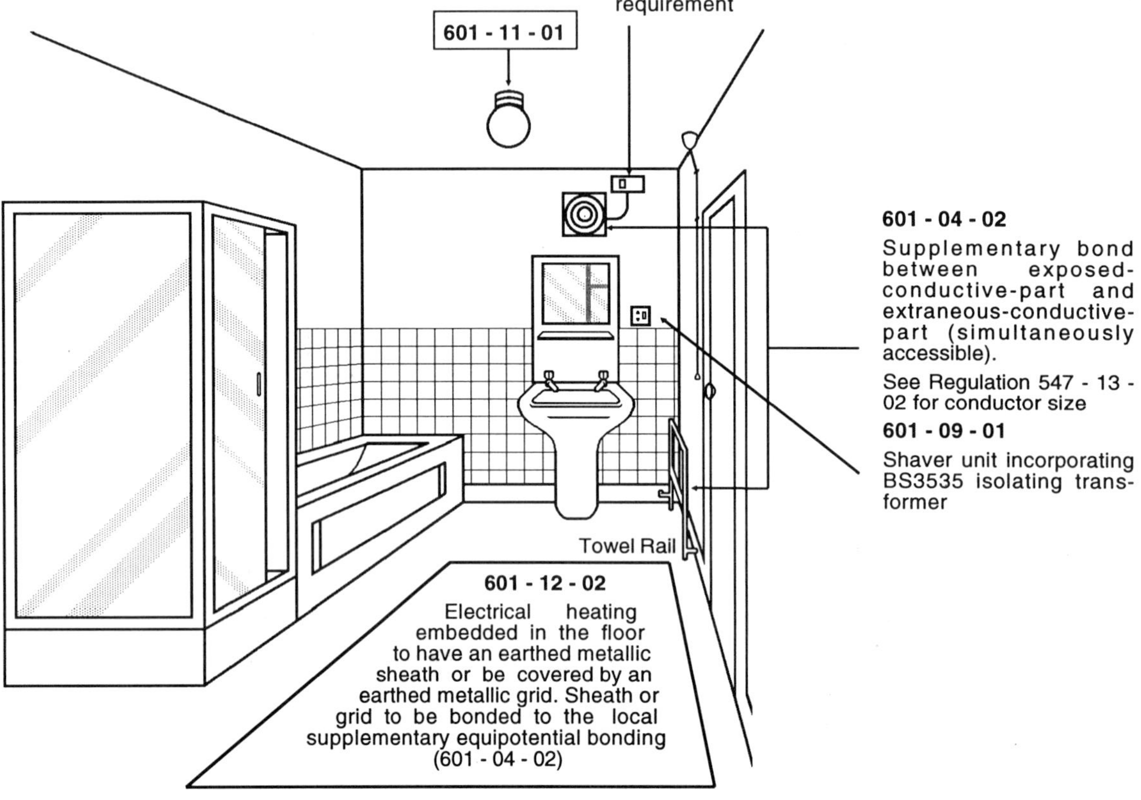

Tests

These must be carried out to ensure supplementary equipotential bonding is connected to the taps or bath or basin

601 - 04 - 02

Irrespective of impedance test results, this Regulation requires supplementary equipotential bonding as follows:-

(1) between simultaneously accessible exposed-conductive-parts of equipment, e.g. heaters, electric towel rails, electric showers, etc.

(2) between exposed-conductive-parts and simultaneously accessible extraneous-conductive-parts

(3) between simultaneously accessible extraneous-conductive-parts

Figure D5

Extra Low Voltage Systems Supplied from More Than One Source (Regulation 551 - 3)

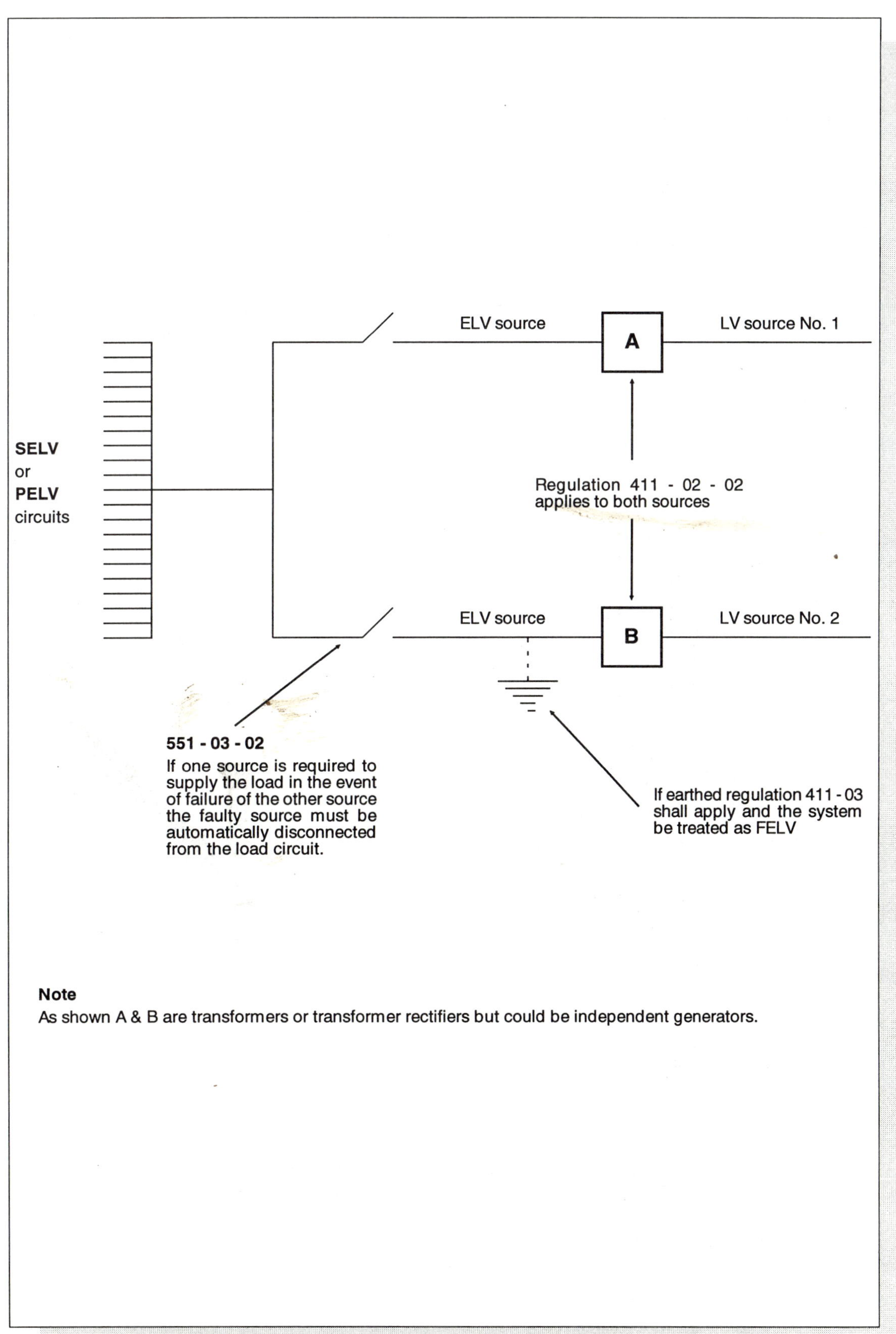

Figure D6

Protection of all Generator Supplies

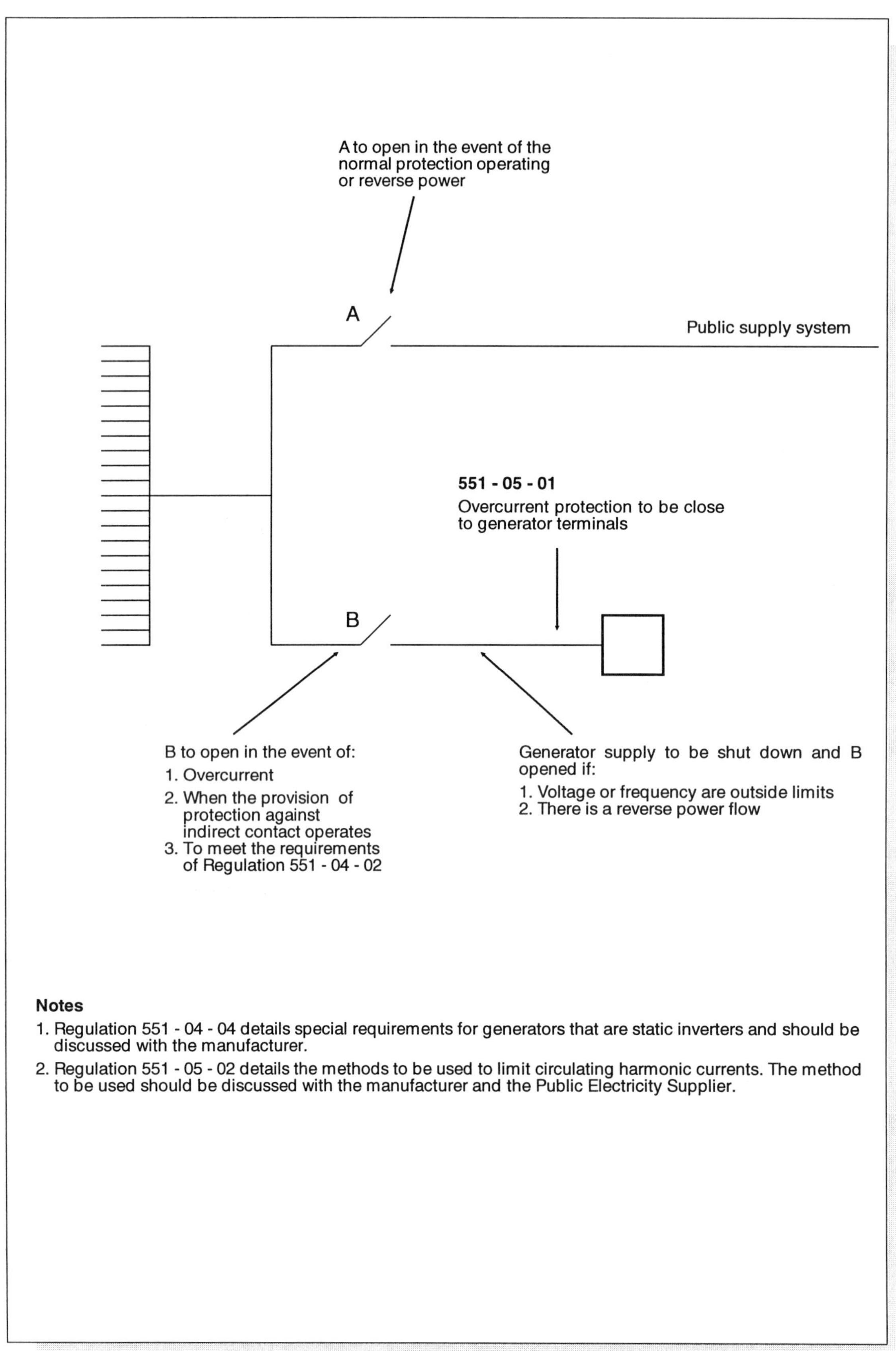

Notes

1. Regulation 551 - 04 - 04 details special requirements for generators that are static inverters and should be discussed with the manufacturer.
2. Regulation 551 - 05 - 02 details the methods to be used to limit circulating harmonic currents. The method to be used should be discussed with the manufacturer and the Public Electricity Supplier.

Figure D7

Generators Providing a Switched Alternative to the Public Supply (Standby Supply)

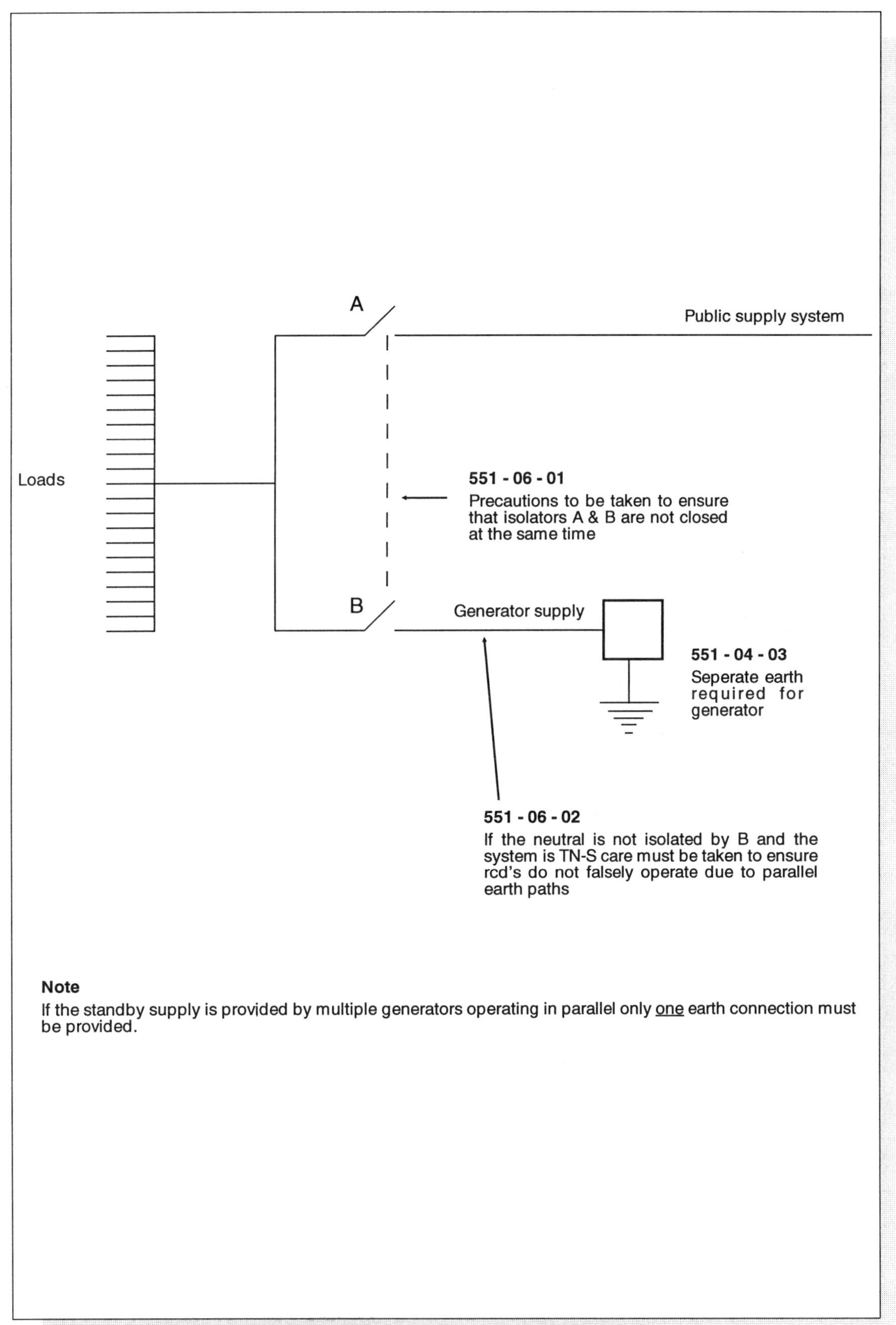

Figure D8

Generators That May Operate in Parallel with the Public Supply System

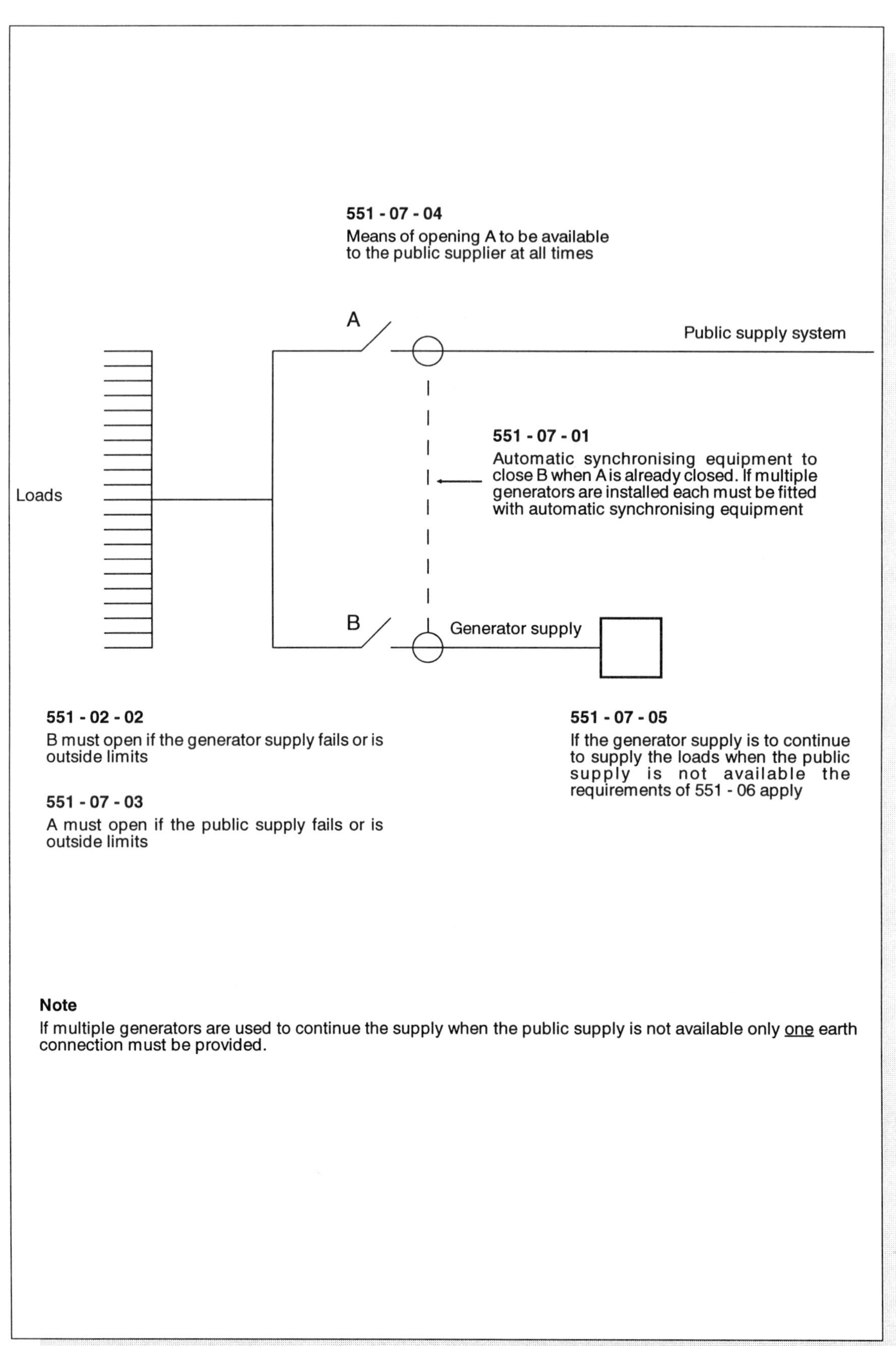

Note

If multiple generators are used to continue the supply when the public supply is not available only one earth connection must be provided.

E

Further Information

Further Information

The various British Standards and Codes of Practice referred to in the Regulations are listed in Appendix 1 of BS 7671:1992. British Standards concerned with Fire Alarms and Emergency Lighting are given in this Section of the Handbook.

Many of the relevant Statutory Regulations and their associated Memoranda are listed in Appendix 2 of BS 7671:1992. This section of the Handbook details more publications which can be of assistance to installation designers and others.

BS 7671: 1992 Amendment 1 incorporated the CENELEC HD 384 series of harmonisation documents listed in the Preface of that BS.

British Standards Associated with Fire Alarm Systems

BS1635 Graphical Symbols and Abbreviations for Fire Protection Drawings
BS5445 Specifications for Components of Automatic Fire Detection Systems
Part 1 – Introduction
Part 5 – Heat Sensitive Detectors
Part 7 – Point-type Smoke Detectors
Part 8 - High Temperature Heat Detectors
Part 9 - Methods of Test of Sensitivity to Fire
BS5446 Specification for Components of Automatic Fire Alarm Systems for Residential Premises
Part 1 – Point-type Smoke Detectors
BS5839 Fire Detection and Alarm Systems in Buildings
Part 1 – Code of Practice for System Design, Installation & Servicing
Part 2 – Specification for Manual Call Points
Part 3 – Specification for Automatic Release Mechanisms for Certain Fire Protection Equipment
Part 4 – Specification for Control and Indicating Equipment
Part 5 – Specification for Optical Beam Smoke Detectors

British Standards and Codes of Practice Associated with Emergency Lighting

BS764 Automatic Change-over Contactors
BS5266 Emergency Lighting
Part 1 – Code of Practice for Premises other than Cinemas
Part 2 – will replace CP1007 – Maintained Lighting for Cinemas
Part 3 – Small Power Relays
CP1007 Maintained Lighting for Cinemas

Statutory Regulations, Associated Memoranda and Other Information

Appendix 2 of BS 7671: 1992 lists a number of relevant statutory regulations and their associated memoranda, whilst Appendix 1 lists British Standards and Codes of Practice referred to in the Regulations. There are various other publications which can be of assistance to installation designers and others, these include:

Association of British Theatre Technicians, 47 Bermondsey Street, London SE1 3XT, Tel: 0171-403 3778 Fax: 0171-378 6170

Safe Sound - Guide to Electrical Safety for Musicians

The Brewers Society, 42 Portman Square, W1H OBB, 0171-486 4831

Code of Practice for Electrical Safety in Beer Dispense in Licenced Premises (1990)

British Approvals Service for Electrical Equipment in Flammable Atmospheres. (BASEEFA)

BASEEFA LIST 1993: Certified and approved electrical equipment. 1989
BASEEFA technical rules new series (TRNS) No.1-14

A complete list of BASEEFA publications is available from BASEEFA, Health and Safety Executive, Harpur Hill, Buxton, Derbyshire SK17 9JN

Building Services Research and Information Association, Old Bracknell Lane West, Bracknell, Berks RG12 4AH, 01344 426511

Application guide 1/81 – Locating Fire Alarm Sounders for Audibility
Technical Note 6/76 – Approximation of the Electrical Load of Lighting Installations

The Chartered Institution of Building Services Engineers, Delta House, 222 Balham High Road, London, SW12 9BS, 0181-675 5211

CIBSE Guides
Code for Interior Lighting
Lighting in Hostile and Hazardous Conditions
LG1: The Industrial Environment
LG2: Hospitals and Health Care Buildings
LG3: Areas for Visual Display Terminals
LG4: Sports
LG5: Lecture, Teaching and Conference Rooms
LG6: The Outdoor Environment
LG7: Lights for Offices
LG8: Museums and Art Galleries
TM12: Emergency Lighting
TM16: Fire Precautions – Sources of Information on Legal and other Services

Prefix LG is for a lighting guide. A full list of all CIBSE publications is obtainable from the CIBSE bookshop at the above address.

Church House Bookshop, Great Smith Street, Westminster, London SW1T 9NZ, 0171-222 5520

Lighting and Wiring of Churches

Copper Development Association, Potters Bar, Herts. 01707 650711

Copper for Busbars
Copper for Earthing
Copper for Electrical Purposes

The Electrical Contractors' Association Security Group, ESCA House, 34 Palace Court, Bayswater, London W2 4HY, 0171-229 1266 or
The Electrical Contractors' Association of Scotland Security Group, Bush House, Bush Estate, Midlothian, EH26 0SB, 0131-445 5577

Code of Practice for Security Installations
Code of Practice for Controlled Access Installations
Code of Practice for CCTV Installations
Code of Practice for Security Installations in all Type of Vehicles/Boats excluding ships

The Electricity Association, 30 Millbank, London SW1P 4RD, 0171-834 2333

The Electricity Association publish a great many well illustrated publications, which are always being added to. They include:

Information on lighting in all aspects and environments
Guidance on heat pumps, air conditioning etc.
Data sheets and case histories on a wide variety of industrial processes
Booklets on electric space and water heating for industrial, commercial and industrial applications, process heating, furnaces, etc., materials handling and electric vehicles

Engineering Recommendations

G5/3 (1976)	Limits for harmonics in the U.K. electricity systems
G12/2 (1982)	National Code of Practice on the application of protective multiple earthing on low voltage networks
G15 (1971)	Standardisation of electricity supply arrangements in industrialised component buildings
G16 (1972)	Standard specification for a new single phase service termination/metering up to 100A
G22/2 (1984)	Superimposed signals on public electricity supply networks
G59 (1985)	Recommendations for the connection of private generating plant to the electricity boards' distribution systems
P9 (1963)	Supply to welding plant and 1984 Amendment regarding d.c. welders
P22 (1981)	Procedure for advising customers on the unsuitability of water pipes for use as earth electrodes
P26 (1985)	The short circuit characteristics of electricity boards' low voltage distribution networks and the co-ordination of overcurrent protective devices of 230V supplies up to 100A
P25 (1984)	The estimation of the maximum prospective short circuit for three-phase 415V supplies
P28	Electric motors - starting conditions

ACE Reports

ACE 7 (1963)	Supply to welding plant
ACE 15 (1970)	Harmonic distortion caused by converter equipment
ACE 22 (1969)	Survey on the amplitude and frequency of occurrence of surges on M.V. systems
ACE 41 (1975)	Guide to an ERA Code of Practice for the avoidance of interference with electronic instrumentation and systems
ACE 73 (1969)	Limits for harmonics in the U.K. electricity supply systems

Employment Department, Moorfoot, Sheffield, S14 PQ 0114 275 3275

Safety Recommendations for on-the-job Training in Electrical Skills

Fire Protection Association, 140 Aldersgate Street, EC1A 4HX, 0171-606 3757

Automatic Fire Alarm Installations for Protection of Property. 1985
Information on static electricity fire risks
Various booklets on electrical fire hazards

Health and Safety Executive, Rose Court, 2 Southwark Bridge, London, SE1 9HS, 0171 717 600

Publications can be obtained from HSE Books, PO Box 1999, Sudbury, Suffolk CO10 6FS, 01787 881165.

As mentioned in the introduction, HSE have a wide range of explanatory and advisory booklets and pamphlets in addition to the statutory publications. Those listed below are in the order in which they appear in the HSE "Publications in Series List", and are of particular relevance to electrical contractors, but the full list should be consulted for further information. The full list is published yearly and a subscription service for information on new or amended publications is available from the HSE.

AS17	Electricity on the Farm
CS2	Storage of highly flammable liquids
GS6(Rev.)	Avoidance of danger from overhead electrical lines
GS23	Electrical safety in schools
GS24	Electricity on construction sites
GS27	Protection against electric shock
GS34	Electrical safety in departments of electrical engineering
GS37	Flexible leads, plugs, sockets, etc.
GS38(Rev.)	Electrical test equipment for use by electricians
HS(G)13	Electrical testing: safety in electrical testing
HS(G)22	Electrical apparatus for use in potentially explosive atmospheres
HS(G)38	Lighting at work
HS(G)41	Petrol Filling Stations: Construction and operation
HS(G)47	Avoiding danger from underground services
HS(R)25	Memorandum of Guidance on Electricity at Work Regulations
IND(G)56(P)	Flammable Liquids on Construction Sites
PM32(Rev.)	Safe use of portable electrical apparatus (electrical safety)
PM37	Electrical installations in motor vehicle repair premises and amendment sheet
PM38	Selection and use of electric handlamps
PM53	Emergency private generation: electrical safety
PM64	Electrical safety in arc welding
SS2	Safe Use of Ladders
SS3	General Access Scaffolds
SS6	Use of portable electric tools and equipment on construction sites
SS7	Avoiding danger from buried services
SS8	Safety in excavations.
SS10	Tower scaffolds
SS11	Safe Use of Propane and LPG cylinders

Occasional Papers

OP10	Safety of Electrical Distribution Systems on Factory Premises

Unnumbered Reports

–	Programmable electronic systems in safety related applications: an introductory guide. 1987
–	Programmable electronic systems in safety related applications: general technical guidelines. 1987

Her Majesty's Stationery Office, 49 High Holborn, London WC1, 0171-873 9090

Electric Lighting (Clauses) Act, 1899
Electricity Supply Acts, 1882 to 1928
Electricity Act, 1947
Electricity Supply (Meters) Act, 1936, and Permitted Alterations Act 1958
Highly Flammable Liquids & Liquified Petroleum Gas Regulations 1972
Gas Safety Regulations 1988
List of Certificates for Intrinsically Safe and Approved Electrical Apparatus for use in certain specified atmospheres
The Electricity Supply Regulations 1988 as amended
The Electricity at Work Regulations 1989
Managing Health and Safety in Construction: Principles and Application to Main Contractor/ Sub-Contractor Projects. 1987 (HMSO 0118839896)

Home Office Fire Department, Headquarters, Queen Anne's Gate, SW1H 9AT, 0171-273 2427

Fire Safety Management in Hotels and Boarding Houses
Draft Guide to Fire Precautions in Hospitals
Guide to Fire Precautions in Existing Places of Entertainment and Like Premises
Guide to Fire Precautions in Existing Hotels & Boarding Houses that Require a Fire Certificate

Institution of Electrical Engineers, Savoy Place, London WC2R 0BL, 0171-240 1871

IEE Guidance Notes

1. Selection & Erection
2. Isolation & Switching
3. Inspection & Testing
4. Protection Against Fire
5. Protection Against Electric Shock
6. Protection Against Overcurrent
7. Special Locations

On-Site Guide

The Lighting Industry Federation Ltd., Swan House, 207 Balham High Road, London SW17 7BQ, 0181-675 5432

Energy Managers Lighting Handbook
(PCBs) Lighting Guide
Better Lighting at Work

LIF3	Factfinder	Lamp Guide
LIF5	Factfinder	Benefits of Certification [ICEL]
LIF6	Factfinder	Hazardous Area Lighting

The Loss Prevention Council, 140 Aldersgate Street, London EC1A 4HY, 0171-606 1050

LPC Rules for Automatic Sprinkler Installations
Plus a range of other publications

National Engineering Specification Ltd., Southgate Chambers, 37/39 Southgate St., Winchester SO23 9EH, 01962-842058

The National Engineering Specification

National Inspection Council for Electrical Installation Contracting, Vintage House, 36/37 Albert Embankment, London SE1 7UJ, 0171-582 7746

Snags and Solutions

NHS Estates, Department of Health, 1 Trevelyan Square, Boar Lane, Leeds LS1 6AE, 0113 254 700

Hospital Technical Memoranda
No. 1 – Anti-static precautions: rubber, plastics and fabrics
No. 2 – Anti-static precautions, flooring in anaesthetising areas
No. 82 – Fire Alarm And Detection Sytems
No. 2007 – Electrical Services: supply and distribution
No. 2011 – Emergency Electrical Services
No. 2014 – Abatement of Electrical Interference
No. 2015 – Patient/nurse call systems
No. 2017 – Engineering Installations: Commissioning and associated activities
No. 2020 – Electrical Safety Standard – LV Systems
No. 2021 – Electrical Safety Standard – HV Systems

Office of Telecommunications - OFTEL, Export House, 50 Ludgate Hill, EC4M 7JJ 2HQ, 071-634 8764

Telecommunication Wiring in Business Premises and Houses

Royal Society for the Prevention of Accidents, Occupational Safety Section, Head Office, Cannon House, Priory Queensway, Birmingham B4 6BS, 0121-200 2461

The Safe Use of Electricity
Range of safety publications, notices and training aids

Telecommuncations Users' Association, 48 Percy Road, London, N12 8BU 0181-445 0996

Telecom Users' Handbook

F

Index to Handbook

Index to Handbook

References are to page numbers in the Handbook